Springer Climate

Series Editor

John Dodson, Menai, Australia

Springer Climate is an interdisciplinary book series dedicated to climate research. This includes climatology, climate change impacts, climate change management, climate change policy, regional climate, climate monitoring and modeling, palaeoclimatology etc. The series hosts high quality research monographs and edited volumes on Climate, and is crucial reading material for Researchers and students in the field, but also policy makers, and industries dealing with climatic issues. Springer Climate books are all peer-reviewed by specialists (see Editorial Advisory board). If you wish to submit a book project to this series, please contact your Publisher (elodie.tronche@springer.com).

More information about this series at http://www.springer.com/series/11741

Shouraseni Sen Roy

Linking Gender to Climate Change Impacts in the Global South

Springer

Shouraseni Sen Roy
Department of Geography and Regional Studies
University of Miami
Coral Gables, FL, USA

ISSN 2352-0698   ISSN 2352-0701   (electronic)
Springer Climate
ISBN 978-3-319-75776-6   ISBN 978-3-319-75777-3   (eBook)
https://doi.org/10.1007/978-3-319-75777-3

Library of Congress Control Number: 2018935962

© Springer International Publishing AG, part of Springer Nature 2018
This work is subject to copyright. All rights are reserved by the Publisher, whether the whole or part of
the material is concerned, specifically the rights of translation, reprinting, reuse of illustrations, recitation,
broadcasting, reproduction on microfilms or in any other physical way, and transmission or information
storage and retrieval, electronic adaptation, computer software, or by similar or dissimilar methodology
now known or hereafter developed.
The use of general descriptive names, registered names, trademarks, service marks, etc. in this publication
does not imply, even in the absence of a specific statement, that such names are exempt from the relevant
protective laws and regulations and therefore free for general use.
The publisher, the authors and the editors are safe to assume that the advice and information in this book
are believed to be true and accurate at the date of publication. Neither the publisher nor the authors or the
editors give a warranty, express or implied, with respect to the material contained herein or for any errors
or omissions that may have been made. The publisher remains neutral with regard to jurisdictional claims
in published maps and institutional affiliations.

Printed on acid-free paper

This Springer imprint is published by the registered company Springer International Publishing AG part
of Springer Nature.
The registered company address is: Gewerbestrasse 11, 6330 Cham, Switzerland

To my husband, Oliver Martin, for his unrelenting support and always being there.

Preface

The Global South is broadly defined as the nations of Africa, Central and South America, and most of Asia including the Middle East. It encompasses at least 150 of the world's 184 recognized states, with many markedly less developed or with severely limited resources. It is widely validated that the impacts of climate change—including heat waves, flooding, drought, and famine—will be felt most strongly by poor communities in the less-developed countries of the Global South, where impacts are exacerbated by poverty and the lack of infrastructure. In general, the challenges and impacts of projected and already-occurring global warming are broadly similar throughout the Global South. In view of the differential impacts of climate change on men versus women, some of the main questions addressed in the book are:

- How does climate change affect access to regional opportunities and resources in the form of education, participation in decision-making processes, food security, and health resources?
- What kinds of obstacles are created by climate change processes that affect women more distinctly than men?

Variability in climatic conditions impacts human and physical systems at different geographic scales and is further complicated by local environmental factors and topography, which affect the vulnerability of resident populations. Negative impacts of climate change are also experienced within relatively small time periods in the form of natural hazards—landslides, floods, hurricanes, and in the long run by the gradual degradation of the immediate environment (damaging effects of long-term increases in temperatures and the frequency of extreme weather events). Adverse impacts of climate change are already evident on agricultural productivity and food security in response to worsening droughts, increasingly vulnerable biodiversity and ecosystems, stressed water resources, human health problems such as infectious diseases spread across newer territories and settlements, migration patterns, energy, transport, and industry. These detrimental impacts of climate change, especially in developing countries, often affect women more than men. A clear majority of the world's 1.5 billion people living on $1 a day or less are women (UNFPA, 2009), who rely most heavily on those natural resources susceptible to climate change. In many cases, women also face greater social, economic, and political challenges than males, which limit their coping capacity. Some of the critical issues can be summarized as follows:

- In rural areas women are usually responsible for collecting water for the families as well as fuel wood for cooking and heating. Specifically, in almost two-thirds of developing countries, women and girls are primarily responsible for obtaining water for their households and communities (UNDP, 2009). They often have to walk longer distances to collect water and fuel wood due to lengthening and intensified droughts resulting from deforestation and resource depletion. As distances to water sources lengthen due to increasing droughts, they become more vulnerable to violence and to the vagaries of harsh weather conditions, such as high temperatures. Not surprisingly, this detracts from the time they can spend getting and completing their education.
- Unequal access to resources and decision-making processes, along with limited mobility, exposes women in rural areas to more of the negative effects of climate change. Often, in view of the traditionally limited role of women in decision-making processes at the household, village, and national levels in most cultures, their needs, interests, and constraints are poorly reflected in policy-making processes and governmental programs aimed at poverty reduction, food security, and environmental sustainability (FAO, 2014).
- Mortality associated with indoor smoke from burning solid fuel amounts to nearly 4.3 million deaths per year, with women and children in rural areas at greatest risk (WHO, 2014). About 25% of these fatalities occur in India, where women and children spend more time in homes that burn fuel wood and coal for cooking and heating purposes (UNDP, 2011). Furthermore, it has been noted that some 80% of rural women in Asia are affected by the growing shortage of firewood (UNDP, 2009).
- A greater proportion of women in poor countries engage in subsistence farming and water collection, exposing them more adversely to the effects of environmental degradation in the form of food shortages and malnutrition (UNFPA, 2009).

In this context, two of the major goals identified in the recently adopted Sustainable Development Goals (SDG) by the United Nations (UN) include taking urgent action to combat climate change and its impacts, and achieve gender equality and empower all women and girls (UN, 2015).

Some of the main aspects of these goals include strengthening resilience and adaptive capacity to climate-related hazards, improving education and awareness about climate change-related issues, and incorporating climate change issues in national and local level policies. Through the gender empowerment goal, the UN aims to eliminate all forms of violence against women and girls, harmful practices against women and girls, and ensure full and effective participation and equal opportunities at all levels of decision making and political processes. Thus there is greater emphasis on gender mainstreaming methodology, which consists of integrating a gendered approach into development and environmental efforts

Preface

(UNCCD, 2011). It includes women's participation in existing strategies and programs. Furthermore, there is greater emphasis on gender-related issues in climate change dialogues, programs, and projects.

The disproportionate burden of long-term impacts of climate change on women has recently been highlighted in reports published by various UN agencies, including UNDP, FAO, WHO, and UNESCO. These reports focus on specific impacts of climate change on women in various regions of the Global South, highlighting issues such as human health, agriculture and food security, and indoor pollution. They further highlight the importance of incorporating the differential impacts of long-term climate change on women, and thereby encourage the participation of women in decision-making processes at the local level. This is particularly significant in the case of South Asia, which is undergoing a massive transformation in its socio-ecological sphere related to climate change while traditional patriarchal societies face strong challenges to their gendered power structures. Therefore, this book highlights the gendered differences in the impacts of climate change in the Global South examined under the major themes of health, water, conflicts, extreme weather, food security and nutrition, and changing urban social contexts. It consists of a comprehensive analysis of the disproportionately negative impacts of climate change on women and girls in the Global South in the context of the issues highlighted above.

Coral Gables, FL, USA   Shouraseni Sen Roy

Acknowledgments

This book has been made possible thanks to a lot of unconditional support and encouragement from my family, friends, colleagues, and students throughout the writing of the book. I would like to start out by thanking my father, who refused to give up on me, and my mother for getting me started on this path.

I would like to extend my appreciation to the University of Miami for the research leaves and research grants to enable all of the research for this book. Thank you to everyone at the UM Writing Center and Richter Library for providing the amazing service and resources on campus. I would like to acknowledge Richard Grant for his advice about embarking on this big project and his valuable feedback for my book proposal and the publication process. I could not have completed all the data analysis without the help of Chris Hanson and Zella Conyers. I am thankful to many of my colleagues at the University of Miami, Miguel Kanai, BLOG, and others. I am also grateful to my mentor, Robert Balling Jr., for his continued mentoring and introducing me to academic publishing.

I am also grateful for the Ann U. White Grant from American Association of Geographers to support my field work in India.

I would like to thank my family in India for being there for me always and my mother-in-law for her free publicity about my work. Finally, a big thank you to my husband for supporting me through all the long conversations about the different ideas in the book, giving honest feedback, and believing in the whole process.

Contents

1 Climate Change in the Global South: Trends and Spatial Patterns . . . 1
Introduction 1
How Much Has the Earth Warmed? 5
Has It Become More Dry or Wet? 8
Is There More Air Pollution? 10
Are There More Intense and Frequent Extreme Weather Events? 12
Has the Climate Become More Variable or Unpredictable? 13
What Are the Model Predictions for Future Climate? 15
Conclusions 19
References 20

**2 Spatial Patterns of Gender Inequalities/Inequities Across
the Global South** 27
Introduction 27
Human Development Index (HDI) 31
Gender Development Index (GDI) 33
Gender Inequality Index (GII) 34
Global Gender Gap Index (GGGI) 36
Education 39
Empowerment 42
Health and Survival 45
Climate Vulnerability Index 47
References 51

3 Health 53
Introduction 53
Infectious Diseases 57
    Malaria 58
    Dengue 62
Extremes 65
    Heat Waves 66
Climate Variability 67

xiii

Global Patterns of Female Health Indicators in Relation to Climate
Change and Vulnerability Index 69
References. 71

4 Water 75
Introduction. 75
Water Scarcity. 78
Water Surplus 81
Water Quality 83
Waterborne Diseases. 86
References. 89

5 Climate Refugees. 93
Introduction. 93
Sea Level Rise. 97
Extreme Weather Events. 101
Food Insecurity 103
Conflicts 108
References. 112

6 Resilient and Sustainable Cities. 117
Urban Heat Island. 120
Land Use and Land Cover Changes (LULCC). 123
Air Pollution 125
Food Insecurity 129
Personal Security in the Context of Changing Social Landscapes 132
References. 134

**7 The Three "E" Approach to Gender Mainstreaming in Climate
Change: *Enumeration, Education, Empowerment*** 139
Introduction. 139
Gender Mainstreaming 141
Enumeration 142
Education 144
Empowerment. 146
Conclusions. 147
References. 148

Index. 149

Chapter 1
Climate Change in the Global South: Trends and Spatial Patterns

> The warnings about global warming have been extremely clear for a long time. We are facing a global climate crisis. It is deepening. We are entering a period of consequences.
> —Al Gore

Sustainable Development Goal 13: "Take urgent action to combat climate change and its impacts"(United Nations 2016).

Introduction

In a recently published commentary piece in Nature journal, the authors called for "ditching" the global warming target of 2 °C above pre-industrial levels that was signed by over 200 countries in the 2009 Copenhagen Summit (United Nations 2016). They argue that this goal is unachievable and impractical. They suggest that instead of monitoring the global average planetary temperatures, certain "vital signs" should be tracked. These vital signs include concentrations of CO_2 and other greenhouse gases, pollutants such as methane and soot that have local and regional implications, ocean heat content, and high-latitude temperatures. Since it was published in October 2014, this article received widespread coverage particularly in view of the then upcoming COP21 conference in Paris in 2015. It highlighted the issue of climate change and associated impacts, which has led to debates both for and against the authors' viewpoints. Several scientists, including Stefan Rahmstorf at Real Climate, William Hare at Climate Analytics, David Roberts at Grist, and Jonathan Koomey at EcoWatch strongly urged for more urgent action toward the 2 °C target (Koomey 2015). The net positive result of all of this discussion is that it has brought the climate change debate back to the forefront. It is now widely mentioned in political speeches either in the form of denial or call for action. There is at least one climate change related story daily across all leading news websites. In this regard a major movement started by The Guardian newspaper in collaboration with 350.org, an activist group, is titled "Keep it in the Ground" campaign. The

---

© Springer International Publishing AG, part of Springer Nature 2018
S. Sen Roy, *Linking Gender to Climate Change Impacts in the Global South*,
Springer Climate, https://doi.org/10.1007/978-3-319-75777-3_1

movement has been gaining momentum through the support of major universities and celebrities, while also encouraging divestment of their assets from oil, gas, and coal companies. Some of the major supporters of this campaign include the Rockefeller Brothers Fund, universities such as Stanford, Glasgow, and Australian National Universities, and the British Medical Association. The discussion about climate change was further invigorated by the latest statements on climate change through the papal encyclical in June 2015 by Pope Francis. He emphasized for an urgent need to develop policies to reduce the emission levels of greenhouse gases, and substitute fossil fuels by developing renewable sources of energy.

Given the renewed interest in climate change related processes and impacts, it is important to first look at the facts from the recent report on climate change by the Intergovernmental Panel on Climate Change (IPCC). During the past few decades, research on climate change has expanded significantly due largely to questions regarding the buildup of greenhouse gases in the atmosphere. The main findings concerning long-term trends in climate change from the recently published Fifth Assessment Report of the IPCC (Stocker et al. 2013) include the following:

- It has been widely validated that the entire globe has experienced warming, although with substantial decadal and inter-annual variability. The average global linear trend for combined land and ocean surface temperatures indicate a warming of 0.85 (0.65–1.06)°C between 1880 and 2012. For the most recent decade, 2003–2012, this trend is 0.78 (0.72–0.85)°C.
- There is significant evidence demonstrating the very likely decrease in the number of cold days and nights, as well as an increase in the number of warm days and nights on a global scale. These findings are further intensified by the higher frequency of heat waves in parts of Europe, Asia, and Australia.
- In addition, increases in number of heavy precipitation events have been recorded in many regions of the world.

Overall, the greatest negative impact of climate change will be on the poor regions of the world, concentrated mainly in the Global South. However, significant uncertainty exists concerning regional variations in climate change impact due to inadequate long-term data records as well as insufficient coverage in these countries. Additionally, the adverse social and economic impacts of climate change are further aggravated by widespread poverty and lack of adequate public infrastructure. Therefore, a better understanding of the impacts of global environmental change will require an integrated international effort to collect, synthesize and analyze the pertinent data in order to determine the most important actions (Stern 2007). Specifically, there is uncertainty as to how local populations and their ecosystems will be affected by and adapt to these changing conditions at various spatial scales, particularly in the vulnerable regions of the Global South.

The Global South, consisting of Africa, Southeast Asia, the Middle East, Mexico, Central and South America, also overlaps with the less developed and poorer regions of the world (Fig. 1.1). It comprises regions at different levels of development along with the highest population density and lowest per capita income levels. Some regions in the Global South are also behind in technology, with massive gaps in

Introduction

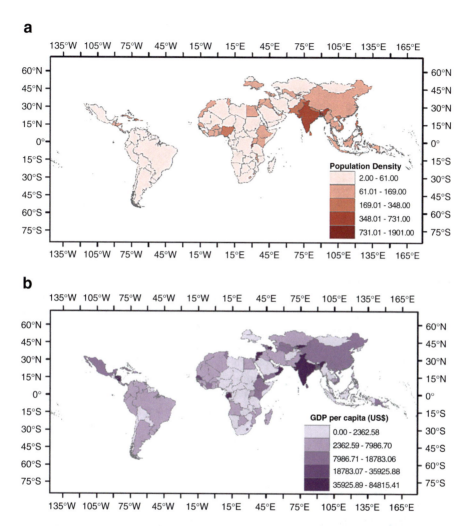

Fig. 1.1 Spatial variations in the Global South (**a**) Population Density; (**b**) GDP Per Capita (US$)

basic infrastructure for their rapidly increasing populations. Furthermore, there is substantial instability in the political, economic, and social spheres, which will magnify the future impacts of climate change such as severe weather events (including droughts, desertification, floods, and heat waves). This widespread instability will result in displacement of populations, ethnic and regional conflicts, hunger, and infectious diseases. As shown in Fig. 1.1 the majority of the Global South is concentrated in the tropics and subtropics on both sides of the equator. Eight of the ten most populous countries in the world are located in the Global South, which is projected to increase to nine out of ten countries by 2050 (Fig. 1.1a) (Population Reference Bureau 2015). Additionally, the top ten countries with the highest fertility rate are all located in the Global South, specifically in Africa, along with some of the lowest GDP per capita.

However, this region also offers many opportunities in the form of emerging technologies, natural resources, and innovative approaches to dealing with impending climate change. For instance, sea level rise is considered to be one of the most menacing effects of global warming. This would affect a major proportion of the population in the Global South, due to the high density of population concentrated in the low lying coastal areas in the larger continents. As a result of sea level rise some of the smaller islands in the Caribbean and Pacific risk disappearing entirely. Sea level rise would also have massive detrimental impacts for the small island nations whose economies greatly depend on tourism. It is estimated that 25% of the population across the African continent will be affected by sea level rise (Juma 2010). On the other hand, desertification is another major impact of climate change that will affect major areas of the Global South, including the Middle East, Southern Africa, Central and East Asia. There is evidence that the recent conflicts in Darfur were aggravated by desertification (Juma 2010). Conversely widespread flooding in parts of northwestern Pakistan, Bangladesh, the Philippines and other parts of Asia has been caused by unusually extreme heavy precipitation events and hurricanes. This has resulted in cutting off large populations' access to food and relief for days and weeks. Additionally, the adverse impacts of climate change are magnified on vulnerable populations including low income communities, women, and children. Furthermore, the access to limited essential resources, including access to clean drinking water and livelihood opportunities, is compromised to a greater extent for the people living in the Global South. Therefore, it is important to examine regional scale long term trends in climate change related processes and their impacts in the Global South.

The focus of this chapter is to gain an in-depth understanding of long term trends in climate change processes and future predictions across the Global South. According to the latest report of the IPCC, the climate system has experienced increased rates of warming over the last three decades, compared to all decades before 1850. The recent period from 1983 to 2012 was the warmest 30 year period over the last 1400 years in the northern hemisphere. This was followed by the recent announcement by the National Climate Data Center (NCDC) that the average temperatures over land and oceans in 2014 were the highest observed temperatures since record keeping started in 1880. It was also noted that there was substantial inter-annual and decadal variability, with periods of weaker trends or hiatus in warming since 1998. Specifically, the decade of the 2000s has been identified as the warmest decade in instrumental record. These observations are based on measurements from a variety of sources including ground-based station observations, remote sensing (including satellites), and other proxy sources. There is also a general consensus about the aggravating role of anthropogenic activities related to the buildup of greenhouse gases on the resulting long term trends in climatic conditions across the earth's surface.

This warming is not only restricted to the land surface but also extends to the oceans, which have experienced significant warming at various depths during multiple time periods. For decades the oceans have absorbed 90% of the earth's heat generated by greenhouse gas emissions. This warming of the oceans has been identified as one of the main contributors to the total overall heating rate in the climate system. As the oceans slowly start releasing the heat back to the atmosphere it will lead to a greater rate of warming of the earth's atmosphere. These long term trends in the climate system have been further modified by variability at the inter-annual and inter decadal scales.

Despite the overall long term trends in surface and ocean temperatures, there are substantial differences when these trends are examined at the regional scale. For example, the tropics are more vulnerable to the direct and immediate impacts of climate change and its associated variability. Therefore, in the following sections the different aspects of climate change processes and related spatial variations in the Global South are highlighted through some basic questions. These questions are addressed through a review of various scientific publications and relevant case studies from the Global South.

How Much Has the Earth Warmed?

There is widespread consensus about the increasing trend in temperatures throughout the globe. As mentioned above there is a clear positive trend in the long term temperature records across the earth, particularly over the last two decades. Based on the analysis of several datasets the trends in average temperatures are the highest over the Global South, which is densely populated and characterized by relatively lower levels of development such as parts of Africa and Asia. However, based on the overall trends demonstrated from all datasets, it is evident that nearly the entire earth surface, including land and oceans, experienced positive trends over the time period of 1901–2012. The only exceptions occur in parts of the northwestern Atlantic Ocean. Some of the areas of highest positive trends are clustered over northern Africa, interior Asia, eastern part of the South America, and higher latitudes in northern North America and the Southern Ocean.

However, the spatial patterns of the overall trends are more informative at the decadal level trends. The data quality improves significantly since 1951 for all regions, specifically across the Global South. During 1911–1940, the positive trends are mostly concentrated across the western half of the earth surface, including the Atlantic Ocean. The middle period from 1951 to 1980 experienced an overall neutral to negative trend across most of the earth's surface. In general the positive trends were more concentrated over the oceanic areas in the southern hemisphere. However, the trends are mostly reversed for the recent-most period of 1981–2012, during which positive trends are observed in all regions of the earth except the eastern Pacific Ocean. This has been attributed to the multi-decadal oscillation in the northern Pacific Ocean, referred to as the Pacific Decadal Oscillation (PDO), and a weak El Niño Southern Oscillation (ENSO), which will be discussed in a later section. Again, the strongest positive trends are concentrated over Asia and northern Africa.

This trend is further validated by the recent announcement by the National Oceanic and Atmospheric Administration (NOAA) about 2014 being the hottest year on record.

Some of the speific aspects of temeprature patterns that have been resarched in detail in nu mberous scientific studies include diurnal temperature range (DTR), calculated as the difference between the daily maximum and minimum temperatures, as well as the maximum and minimum temperatures separately. A recent comprehensive study at the global scale indicated a declining trend in DTR (Vose et al. 2005). This decline was attributed to a faster rate of increase in minimum temperatures. Subsequent studies indicated a decrease in the DTR since the mid-1980s, followed by an increase in DTR in the twenty-first century (Rohde et al. 2013). These patterns are consistent with the global dimming and brightening trends. Despite varying results from analyzing different datasets, the overall consensus is that both maximum and minimum temperatures over land have increased in excess of 0.1 °C (Hartmann et al. 2013). There are substantial variations at the regional scale. For instance, the analysis of DTR, and maximum and minimum temperatures from 1931–2002 in India revealed increases in both maximum and minimum temperatures despite minimal change in DTR (Sen Roy and Balling 2005).

Box 1.1 Melting Ice Caps of Mt. Kilimanjaro

One of the most well documented long term effects of global warming is the melting of the ice caps on Mount Kilimanjaro. The ice caps here have been regularly mapped since 1912. The graphics above show the rapidly shrinking size of the glaciers on top of Mount Kilimanjaro, with predictions of them disappearing completely over the next few decades. The ice cover on the summit today is only 15% of what it was in 1912. Additionally, a recent study by Thompson et al. (2009) revealed almost 26% of the ice loss occurred in the twenty-first century.

One of the main anthropogenic reasons for this rapid shrinking of the ice cover is massive deforestation occurring on the slopes of Mount Kilimanjaro. Forests are being cleared around the mountains for agriculture, which alters the earth-atmosphere interaction leading to decreased moisture supply in the atmosphere through transpiration from trees. This results in drier conditions and more solar radiation reaching the surface, which further accelerates the rate of evaporation on the surface of the glaciers. Additionally, the top of the mountain is very dry, with hardly any precipitation. Therefore, once the ice melts, the dirty tops of the glacier absorb more solar radiation which in turn accelerates the rate of melting. Other than warmer temperatures on the mountain top, the role of the warmer ocean temperatures in the adjacent Indian Ocean is also said to play a role in the melting of glaciers on the top of this mountain.

One of the main factors affecting trends in land surface temperatures at the local scale is the role of land-use land cover change (LULCC), which includes the urban heat island (UHI) effect and deforestation. It is important to note that the effect of UHI on long term temperatures varies based on the locations that recently have experienced substantial urbanization compared to those which have been urbanized for an extended period of time.

Box 1.2 Global Warming Hiatus?

It is noteworthy however, that a few recent studies published in leading journals identify no significant increasing trends in annual average temperatures despite rising concentrations of greenhouse gases. This hiatus has been mostly identified as beginning in the twenty-first century. Several possible mechanisms for this phenomenon have been proposed, which include a decline in radiative forcing caused by aerosols in the two lowest levels of the atmosphere (stratosphere and troposphere), an increase in stratospheric water vapor, and a solar minimum around 2009 (Solomon et al. 2010, 2011; Kaufmann et al. 2011). Several studies have also indicated this hiatus to be part of a larger natural variability related to La Niña like cooling conditions coupled with the switch to the decadal scale negative phase of the Pacific Decadal Oscillation in 1999 (Meehl et al. 2011; Kosaka and Xie 2013; Trenberth and Fasullo 2013).

However, this was refuted by a recent publication by scientists at the National Oceanic and Atmospheric Administration (NOAA), who argued that when adjustments are made to the data to compensate for problems in the measurement of global temperatures, then there is no slowdown in the warming (Karl et al. 2015). Rather, the results of their updated analyses reveals similar rates of warming in the first 15 years of the twenty-first century as the in last half of the twentieth century.

Additionally, it is also important to take into consideration what type of prevailing land use was replaced by urban land use. According to the World Health Organization (WHO) the urban population in 2014 constituted 54% of the total world population (compared to 34% in 1960), with most of population growth concentrated in the less-developed-world, which is most of the Global South. For instance, cities such as Phnom Penh, Cambodia, Tijuana, Mexico, Marrakesh, Morocco, and Lagos, Nigeria, all of which are located in the Global South, are projected to grow at annual rates of around 4%, effectively doubling their populations within the next 17 years. Some cities in China, such as Shenzhen and Xiamen, will experience annual growth rates of more than 10%, doubling their populations every 7 years. This increase in urban population is accompanied with unplanned urban sprawl and inadequate infrastructure to support the increasing population density. Thus the impacts of UHI effects are magnified in the large urban areas in the Global South. Specific studies examining UHI effect in large metropolitan areas in the

Global South include Beijing (Ren et al. 2007; Yan et al. 2010), Kuala Lumpur (Sham 1987), Gaborone (Jonsson 2004), Ouagadougou (Lindén 2011), and New Delhi (Sen Roy et al. 2011).

On the contrary, the modifying influence of soil moisture in agricultural areas on local climate processes, in some cases offsetting the effects of climate forcings such as greenhouse gases have been well documented (Lobell et al. 2008; Kueppers et al. 2008; Mahmood et al. 2014). This is particularly significant for the agriculturally intensive parts of the Global South, such as northwest India where cooling effect of irrigation on near surface air temperatures and precipitation has been identified (Douglas et al. 2007; Sen Roy et al. 2007; Lobell et al. 2009). The main mechanism is the increased evapotranspiration in the atmosphere as a result of increased irrigation leading to enhanced moisture availability in the atmosphere.

Has It Become More Dry or Wet?

Most discussions about global warming usually focus on rising temperatures. However, the long-term trends, amount, and timing of precipitation patterns across the earth constitute an important aspect of climate change that has significant long-term impacts, such as food security and human health. The analyses of most global datasets show an increasing pattern for average precipitation. More specifically the positive trends were located north of 30°N, while the trends were reversed in the tropics between the 1970s and early 2000s. After this period there has been a positive trend in precipitation patterns. Longer term trends are available for the higher latitudes mainly due to better quality of data records for longer periods of time. Major parts of the Global South such as parts of Africa, Asia, and South America do not have long term records of climate data. There is limited data coverage over most of the Global South during the earlier part of the twentieth century. Additionally, the analysis of snowfall indicated a declining trend over areas where there was an increasing trend in temperatures, such as the Himalayas and parts of East Asia. Since the 1960s, areas of precipitation decrease resulted in a decline in runoffs in various streams in the low and mid latitudes such as the Yellow River Basin in northern China (Piao et al. 2010). On the other hand, increased discharges were found in the Yangtse River in southern China (Piao et al. 2010) and the Amazon basin in South America during recent decades (Espinoza Villar et al. 2009).

Analysis of several long term datasets from Central America and the Caribbean indicate negative trends during 1950–2003 (Neelin et al. 2006). A significant decrease in precipitation over Puerto Rico was caused by the disruption of moisture convergence due to the increased flow of easterly surface winds (Comarazamy and Gonzalez 2011). Additionally, in South America during the twentieth century, the continent exhibited a significant increase in precipitation during the twentieth century in the southern part of southeastern South America and southern Chile (Quintana and Aceituno 2012), while a negative trend in precipitation was observed in the

South Atlantic Convergence Zone (SACZ) continental area and central-southern Chile (Barros et al. 2008).

Another aspect of the hydrological cycle is the rate of evapotranspiration, including pan evaporation. Declining trends were observed over India, China, and Thailand. Some of the possible causes for this decline were attributed to a decrease in surface solar radiation, a decrease in duration of sunshine, an increase in specific humidity, and an increase in cloud cover (Hartmann et al. 2013). More specifically, at the decadal scale, the global evapotranspiration rates over land increased from the early 1980s to late 1990s (Wild et al. 2008; Jung et al. 2010). The decline in the evapotranspiration rates have been attributed to the lack of soil moisture across land areas in the southern hemisphere. Declining trends in evapotranspiration rates were also observed in the Tibetan plateau despite increasing temperatures (Zhang et al. 2007) and northeast India (Jhajharia et al. 2009). Some of the factors affecting evapotranspiration rates include the role of declining wind speeds, land use change, and the lengthening of the growing season in Tibet (Roderick et al. 2007). On the other hand, there has been no significant trend in surface-level specific humidity since 2000, which, when considered in relation to warmer temperatures, implies lower relative humidity.

Box 1.3 Deforestation of the Amazon Rainforests

The Amazon rainforest is the largest rainforest in the world and is known for its unique biodiversity and its role in the earth atmosphere interaction processes. However, these rain forests have been subjected to widespread deforestation, which has had broad impacts on the wider ecosystem. Deforestation in this region mainly started in the 1960s from the building of roads and infrastructure, most of which was illegal. The main causes of deforestation include cattle ranching, large scale agriculture, and logging. Most of the deforestation took place in the 1990s when the gross rate of forest clearance was ~25,000 km^2/year (Achard et al. 2002), which has declined in recent years. The important role of forests as carbon sinks is widely recognized. In a recently published study by Exbrayat and Williams (2015), it is estimated that deforestation in the Amazon has led to an 1.5% increase in levels of CO_2. If the deforestation had not taken place then the rainforests would store an additional 12% of carbon in the vegetation. The satellite image below shows the typical spatial patterns of deforestation in the Amazon rain forests.

Additionally, deforestation in the Amazon rainforests has also accelerated the detrimental effects of climate change processes. It impacts the water and energy fluxes between the forests and the atmosphere through the alteration of the physical properties of the land surface including albedo, evapotranspiration rates, and surface texture (Bonan 2008). Lower evapotranspiration rates lead to lower latent heat flux, accompanied by higher sensible heat flux and thus increased surface temperatures. This is evident from the 0.5 to 0.8 °C increase in monthly mean air temperatures over the last decade of the twentieth century (Quintana-Gomez 1999). The impacts are already visible in the form of declining precipitation, particularly in the dry season over northern Amazonia since the mid-1970s, with no consistent trend in the south (Marengo 2004). There is also evidence of long term shifts in the rainfall patterns across the region, with a significant decrease in rainfall at the end of the wet season (Changnon and Bras 2005). Additionally, in view of the projected rapid decrease in the extent of these forests in future decades, climate model simulations predict increased surface temperatures and decreased precipitation locally. This is demonstrated by the recent severe droughts of 2005 and 2010. As a result of land use change in this region, by 2100 the annual mean surface temperatures are predicted to increase by 0.5 °C accompanied by an annual mean decrease in rainfall of 0.17 mm/day compared to present conditions (Lejeune et al. 2015). Other than the direct impacts on long term climate processes, the effects of deforestation will be seen in increased erosion, loss of biodiversity, decreased agricultural yield, and potential health impacts through the spread of infectious diseases.

Is There More Air Pollution?

One of the main factors leading to the increasing trends in near surface air temperatures is the level of emissions of greenhouse gases (GHGs), which include CO_2, Methane (CH_4), Nitrous Oxide (N_2O), Hydrofluorocarbons, Perfluorocarbons, and Sulphur Hexafluoride. Most of the long term trends are based on global measurements with limited spatial coverage. For instance, the global levels of CO_2 emissions from 1980 have increased by 1.7 ppm/year, and since 2001 they increased by 2.0 ppm/year (Hartmann et al. 2013). There is clear evidence of the burden of fossil fuel combustion and land use change such as deforestation on the increasing levels of CO_2 in the atmosphere (Tans 2009). A global assessment of forest resources conducted by the Food and Agricultural Organization (FAO) in 2010 revealed the largest net loss of forests in South America (Brazil), Africa (Republic of Congo, Zimbabwe, and Tanzania), and most countries in Southeast Asia (FAO 2010). In the case of CH_4, the global levels have decreased from the early 1980s to 1998. After a short period of stabilization from 1999 to 2006, it again began increasing from 2007 to 2011 (Rigby et al. 2008). This recent increase in the levels of CH_4 has been mainly attributed to the abnormally high temperature in the Arctic in 2007, along with above-average precipitation in the topics from 2007 to 2008 (Dlugokencky et al. 2009; Bousquet 2011). Finally, the global average level of another GHG, N_2O, increased by 0.75 ppb/year since the 1970s (Hartmann et al. 2013). The increase in the levels of N_2O are mainly attributed to anthropogenic emission from the use of fertilizers in the northern tropical to mid latitudes and natural emission from soil and ocean upwelling.

In addition to the GHGs, there are other gases and particulate matter such as aerosols, stratospheric water vapor, tropospheric and stratospheric Ozone (O_3), and precursor gases, including nitrogen dioxide (NO_2) and carbon monoxide (CO) that have a clear impact on the long term trends in climate processes driven by local processes. In this context there is substantial evidence that stratospheric O_3 has declined during the 1980s and early 1990s, while remaining constant over the past decade (Hartmann et al. 2013). On the other hand tropospheric O_3, which has many detrimental effects including photochemical smog, exhibited increasing trends in the mid latitude and tropics of the northern hemisphere, including India and China (Schnadt Poberaj et al. 2009). Furthermore, the rapid rate of urbanization has highlighted the increased levels of intense air pollution in large urban areas in the Global South such as Mexico City (Baumgardner et al. 2004), Sao Paulo (de Miranda et al. 2002), Buenos Aires (Mazzeo and Venega 2004), Delhi (O'Shea et al. 2015), Kolkata (Ghose et al. 2005), Hong Kong (Tanner and Law 2003), and Taipei (Li and Lin 2002).

A discussion about air pollution is not complete without a discussion about aerosols, defined as fine suspended particulate matter in the atmosphere. Aerosols can scatter and absorb incoming solar radiation, and thus reduce visibility. Aerosols play an important role in the context of global dimming and brightening depending on the specific properties. In addition due to the short life span of tropospheric aerosols, they tend to have clear regional patterns (Hartmann et al. 2013). There are two types of aerosols: anthropogenic sources of aerosols that are confined to more populated regions in northern hemisphere; and the natural sources of aerosols, mainly volcanic, sea salt, and desert dust, which are more closely related to climate and land use change (Carslaw et al. 2010). Analysis of long-term data on aerosols, conducted from 1973 to 2007, indicates a constant positive rate over most of the Global South, mainly Asia, parts of South America, the Arabian peninsula, Africa, and Australia. For instance, observations from aerosol monitoring stations over India indicated an increase of 2% annually during the last two decades (Krishna Moorthy et al. 2013), which is mostly related to mining and industrial activity over northern India. Additionally, strong positive trends were also observed over oceanic areas adjacent to southern and eastern Asia as well as over most tropical ocean areas (Hartmann et al. 2013). The positive trends observed in the Arabian Peninsula and Saharan outflow region of western Africa were related to dust. Furthermore, there are substantial seasonal differences in the spatial patterns of aerosols, with strong presence of aerosols over the Arabian Peninsula during the spring and summer dry months from March to August, and over southern and western Asia during the dry months of December to May. A recent report by the WHO identified the top three cities in terms of particulate matter in India, with New Delhi at the top of the list (WHO 2015). In addition, 7 million premature deaths annually are linked to air pollution exposure, with most of these deaths concentrated in Southeast Asia and Western Pacific regions. Among these deaths, about 3.3 million deaths, most of who are women and children, are related to indoor air pollution, and 2.6 deaths are related to outdoor air pollution (WHO 2015).

Are There More Intense and Frequent Extreme Weather Events?

One of the main aspects of climate change is the occurrence of extreme events, such as heavy precipitation, extended periods of high or cold temperature days, and severe weather events like hurricanes, tornadoes, floods, and droughts. These events have a major impact in terms of their widespread impact on the society and ecosystems. The trends in extreme maximum and minimum temperatures have increased globally since 1950 (Donat et al. 2013a, b). At the regional level, parts of Eurasia and the Asia Pacific region have experienced doubling of the occurrence of warm nights and halving of the occurrence of cold nights (Choi et al. 2009; Donat et al. 2013a, b). The overall global trends for temperature extremes show a warming trend, particularly for the minimum temperatures since the twentieth century. For instance, there is a positive trend in the frequency of heat waves across most of Asia. However, there is limited confidence in the trends over South America and Africa due to insufficient data and inconsistent methodologies used for defining heat waves. In addition, there is substantial evidence demonstrating the overall increase in the number of warm days and nights, and decrease in the number of cold days and nights across Asia (Hartmann et al. 2013). Furthermore, a recent study noted the increased frequency of heat waves and reduced cold spells in urban areas (Mishra et al. 2015).

Box 1.4 Long Range Cross Continental Transport of Pollutants

With the rapid rate of urbanization and development in the Global South there is growing concern about the long range transportation of pollutants across large distances and international boundaries. In some cases there is evidence of the transfer of aerosols and dust pollutants loaded with heavy metals from industrialized zones across vast oceans, including the Atlantic and the Pacific. For instance, one of the most intensely studied transcontinental transportations is the dust storm that occurred in April 2001 over northern China. The dust storm started on April 5, 2001 in the Sinkian Basin of northwestern China and southern Mongolia. It spread slowly over eastern China and South Korea by April 7. By April 15, a whitish haze from the storm had reached parts of North America from Calgary, Alberta in the north to Arizona in the southwest. Over the next ten days the haze spread across the entire North American continent extending to the east coast, and finally moving over the Atlantic Ocean. Analysis of some of the dust transported by these dust storms indicate high levels of harmful chemicals such as sulfur, heavy metals, and carcinogens (Erel et al. 2006). High concentrations of dust from dust storms cause poor visibility and health issues.

Such long range transport of pollutants has also been observed in the case of the Saharan dust transport from northern Africa to parts of Europe and North America, black soot from biomass burning in the Amazon, as well as the Asian brown cloud observed over the Arabian Sea related to anthropogenic industrial activities over the adjacent Indian subcontinent.

The majority of the research on extreme precipitation focuses on variables derived from daily precipitation extremes. The overall trends in extreme heavy precipitation patterns indicate an increasing trend, with less coherent spatial patterns compared to temperatures. For instance, South America experienced overall increasing trends in the frequency and intensity of heavy precipitation events (Donat et al. 2013a, b). The trends are mixed in Asia with most regions showing increasing trends. In India the trends were mixed, with positive trends in the frequency and intensity of heavy precipitation events over contiguous regions extending from the northwestern Himalayas to the peninsular south. The trends were reversed across the eastern Gangetic plain and northeast Himalayas (Goswami et al. 2006). Recently, there has been more focus on sub-daily scale precipitation events such as at the hourly scale, including India (Sen Roy 2009) and South Africa (Sen Roy and Rouault 2013). Both of these studies indicate positive trend spatially dominating over negative trends spatially.

One of the natural hazards associated with extreme heavy precipitation events is floods. There are a limited number of studies focusing on the trends in the magnitude and frequency of flooding events in the Global South due to limited data availability with regards to river discharges. However, it is noteworthy that floods are a major natural disaster in the Global South, particularly in the rice-growing regions of south and Southeast Asia, as well as flooding caused by tropical cyclones such as Haiyan (2013) and Hagupit (2014) in the Philippines, and Nargis (2008) in Myanmar. About 4 million people needed to be evacuated for Typhoon Haiyan, which killed 6000 people. Cyclone Nargis affected 2.4 million people, killed 84,500 people, and caused 53,800 to go missing. There are no clear long term trends in the frequency of tropical cyclones in different parts of the world.

Similarly there are no clear long term trends in the frequency of droughts at the global level. This is mainly due to the different proxy measures such as soil moisture, drought proxies (PDSI, SPI), and hydrological drought proxies used for analysis. Due to the complex definition of these variables and their inadequate representation of droughts there are no clear long term trends in drought (Hartmann et al. 2013). At the regional level, positive trends were observed in the length of drought in India, parts of South America, and East Asia (Giorgi et al. 2011), and in the frequency and intensity of droughts across West Africa and the Mediterranean region (Hartmann et al. 2013).

Has the Climate Become More Variable or Unpredictable?

One of the dominant forces driving the inter-annual and decadal variability in regional scale climate processes is the role of teleconnections. It is defined as a statistical association between widely separated climate variables at geographically fixed spatial scales. Therefore, a teleconnection pattern is calculated by the correlation between variables at different spatial locations and a climate index (Christensen et al. 2013). There are several prominent teleconnections that affect climate patterns across the earth, which include the North Atlantic Oscillation (NAO), PDO, and ENSO. Overall, teleconnections are the strongest during winter months when the mean circulation is the strongest.

Among these various teleconnections, ENSO is one of the relatively dominant processes determining the climatic variability, in several regions of the Global South. ENSO is defined as a coupled ocean-atmosphere phenomena naturally occurring at the inter-annual time scale over the tropical Pacific. During an El Niño year, the warmer sea-surface temperatures (SST) are observed on the west coast of the South American continent, thus weakening the usual gradient in SSTs across the Pacific. This is accompanied by changes in trade winds, tropical circulation, and precipitation, represented by the mean seal level pressure anomaly difference between Tahiti and Darwin, and is normalized by the long-term mean and standard deviation of the mean sea level pressure difference (Trenberth et al. 2007). It typically occurs every 3–7 years. In this context the modulating role of ENSO on the Indian monsoon is an extensively researched topic, particularly due to the importance of precipitation for the large agricultural sector of this region. The inverse relationship between ENSO (through changes in the Walker circulation) and the Indian monsoon precipitation (resulting in below normal precipitation and in some cases droughts) is well established (Meehl and Arblaster 1998). However, recent studies examining the relationship between ENSO and the Indian summer monsoon have indicated a weakening in the relationship, which can be attributed to the decadal level oscillation in northern Pacific in the form of the PDO (Krishnan and Sugi 2003; Sen Roy 2006). For instance, during 1997–1998, considered to be one of the strongest El Niño years on record, the monsoon rainfall in India was above average. However, during 2002 which was a weak El Niño year, India experienced one of the driest years (Kumar et al. 2006). Additionally model projections, predict further weakening of the ENSO-monsoon relationship under warmer temperatures (Meehl et al. 2007). The role of ENSO has a dipolar mode on the amount of precipitation occurring across the African continent, where precipitation in eastern Africa is in phase with warm ENSO events but negatively correlated with precipitation occurring in Southern Africa (Nicholson and Kim 1997). Additionally, there is also evidence of a greater role of the Atlantic Ocean (during La Niña years) and Indian Ocean (during El Niño years) on African rainfall (Williams and Hanan 2011). For instance the major ENSO event of 1997–1998 caused extreme wet conditions in eastern Africa, while La Niña event of 1999–2000 led to major floods in Mozambique.

Another teleconnection that has a significant influence over Southern South America and South Africa is the Southern Annular Mode (SAM). The SAM is characterized by differences in geopotential heights over the mid-latitudes in the southern hemisphere (Fyfe 2003). The positive phase of SAM is associated with unusually dry conditions over southern South America, and unusually wet conditions over South Africa (Gillett et al. 2006).

What are the Model Predictions for Future Climate? 15

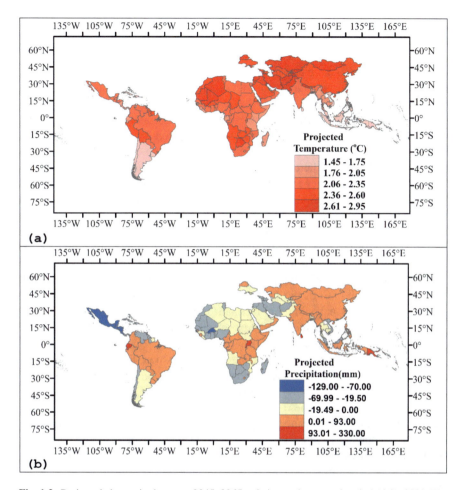

Fig. 1.2 Projected change in the years 2045–2065, relative to the control period 1961–2000. The values reflect the average of range between the tenth and ninetieth percentiles of results from nine general circulation models (GCMs). Values are calculated at the country level from 2-degree gridded data (**a**) Temperature (**b**) Precipitation (Data Source: World Bank 2015)

What Are the Model Predictions for Future Climate?

In order to understand the impacts of the spatial variations of long term climate change, the projected or future predictions of climate change based on climate model simulations need to be taken into consideration. As evidenced in the previous sections, there are substantial differences in the spatial patterns of regional climate change resulting from complex local processes responding differently to changes in global-scale processes. However, climate model simulations are not the most accurate for all climate processes. In general there is better representation and agreement on some of the large scale phenomena such as the monsoons and tropical cyclones which affect large parts of the Global South. The overall projected patterns of average annual temperature and precipitation changes depicted in Fig. 1.2 are derived from climate model simulations for 2045–2065, relative to the control period of 1961–2000. These data are based on the country level data developed by the World Bank and modified from data produced by multi model experiments conducted for the twenty-first century (Meehl et al. 2007).

There is a general consensus among different model simulations about the overall strengthening of monsoonal systems in area and intensity, coupled with weakening monsoon circulation in India and strengthening of the monsoon circulation in East Asia. Additionally, there will be lengthening of monsoon phenomena associated with early or no change in the onset dates but delayed retreat of monsoons in all regions (Christensen et al. 2013). The increase in the intensity of the monsoon implies an increase in the frequency and intensity of extreme precipitation in these regions. Specifically, the monsoonal precipitation over South and East Asia is projected to increase. The projected increases in precipitation associated with the Indian monsoon are related to anthropogenic forcing (May 2011). However, the monsoonal circulation will likely decrease for the Indian monsoon, while it will slightly increase in case of the East Asian monsoon. Historically, the East Asian monsoon has been weakening and has had limited extent in the North since the 1970s. This has resulted in increased droughts in Northern China and flooding in the Yangtze Valley in the South.

One of the main driving forces of the Indian monsoon, ENSO, is predicted to remain as a dominant mode of inter-annual variability in the future, while decreased moisture availability will result in the intensification of precipitation variability. The main processes leading to these changes are related to the more accelerated warming of land surfaces vs, ocean surfaces, resulting in a greater contrast in temperatures. This temperature contrast in turn affects the atmospheric circulation and thus changes in the intensity, timing, and circulation of monsoon systems. Additionally, the warmer temperatures leading to greater availability of moisture in the atmosphere, also leading to increased precipitation and extremes, irrespective of any change in the strength of the monsoonal circulation. Over the next century, climate models predict a decrease in precipitation associated with the North American Monsoon. In the case of the African Monsoon circulation, model simulations indicate a decline in rainfall in the early part of the season followed by an intensification of rains in the late season (Biasutti and Sobel 2009; Seth et al. 2010).

Thus, the general overall prediction for monsoon phenomena across the Global South is that there will be an overall weakening in the circulation over India and West Africa, while the monsoonal circulation will strengthen in case of the East Asian Monsoon and the West African Monsoon. However, as a result of warmer temperatures and increased moisture availability, there will be an increase in extreme heavy precipitation events and intensity in all areas. Additionally, in most areas there will be a delay in monsoon retreat dates.

The projections for the different teleconnections that have a substantial modulating influence on the climate processes, including the monsoons in the Global South, vary spatially. In the case of ENSO, the simulations reveal that the role of ENSO on the inter-annual variability in the future, due to increased moisture availability, will remain a dominant force (Christensen et al. 2013). In the case of the SAM, it is predicted that with the recovery of the O_3 hole, the recently observed positive trends in SAM will be reversed particularly during the late austral spring and summer. This is the period when the O_3 depletion has its greatest impact (Bracegirdle et al. 2013). However, it is also noteworthy that the SAM is not well represented in models primarily due to the difficulty for climate models to accurately represent boundary processes, such as stratosphere-troposphere interaction, ozone chemistry, solar forcing, and atmospheric response to Arctic sea ice loss (Christensen et al. 2013).

However, in the case of tropical cyclones, there is lesser certainty globally about the changes in their frequency of occurrence. The wind speed and precipitation rates associated with tropical cyclones are likely to increase, although with substantial spatial variations. Extreme precipitation associated with tropical cyclones is predicted to increase across Central America, and most of Asia.

At the regional level, the projections are mainly for average surface air temperatures and average precipitation. These projections are based primarily on multi-model ensemble projections from general circulation models. However, the greater uncertainty associated with the regional climate models have been highlighted in latest report from the IPCC. These include natural variability (Deser et al. 2012), aerosol forcing, land use land cover change (Defries et al. 2002; Moss et al. 2010). The relationship between global and regional climate change has been summarized as the regional variations and spatial distribution of heat and moisture that are driven by external forcings (latitude dependent solar radiation, aerosol emission sources, and land-use changes), surface conditions (including land-sea contrast, topography, sea surface temperatures, and soil moisture content), and weather systems and ocean currents that redistribute heat and moisture across the earth's surface. Furthermore, the projected changes in precipitation compared to temperatures are more variable at the regional scale. As a result, the model predictions are more robust for certain regions. For example, the predictions for precipitation patterns in the Central America and Caribbean region are consistent with the past patterns of negative trends observed in this region since 1950. Decreasing trends in precipitation are projected throughout the year in northern Central America and Mexico, the southern Caribbean from October to March, and the entire Caribbean during March to September. In terms of temperature, the greatest warming is projected in this region during the summer months of June, July, and August, specifically over Central America. Additionally, the continuing influence of ENSO on the climate processes in this region remains strong. In South America, the model projections indicate higher temperatures overall, with the maximum changes in southeastern Amazonia during austral winter. A greater frequency of warm nights over most of the continent except parts of Argentina, along with a decrease in the frequency of cold nights across the entire continent is predicted (Marengo et al. 2012). For precipitation, the model projections indicate an increase from October to March in the southern part of Southeast Brazil and the La Plata River basin, the extreme south of Chile, and the northwest coast of South America.

On the other hand, decreased precipitation is projected over eastern Brazil, and central Chile (Christensen et al. 2013).

In the case of the African continent, most model predictions indicate continued warming during the twenty-first century. The Sahara desert is projected to remain very dry, with a probable delay in the West African monsoon and reduced precipitation over South Africa during the austral winter. A recent study consisting of the evaluation of six climate models over East Africa projected no clear trend in mean annual rainfall until the end of the twenty-first century (Conway et al. 2007). Their results indicated an increase in September to November rainfall, but a decrease in March to May rainfall. The predictions for temperature in Central and North Asia are similar to Africa, i.e. a stronger than global mean warming trend. The magnitude of warming is higher during the winter in Northern Asia. Additionally, the predictions show increasing trends in mean and extreme precipitation in both Central and Northern Asia (Seneviratne et al. 2012). Over East Asia, the model predictions project an intensification of the summer monsoon circulation and associated mean and extreme precipitation in a warmer climate (Jiang et al. 2012; Lee et al. 2011). This increase in precipitation is projected for all seasons over East Asia, but only during the summer in Japan. In West Asia, there is general agreement among model simulations about the warming trends, but limited agreement about a declining trend in precipitation. Specifically, the Mediterranean side of the region shows a drying trend. Similar warming trends in temperature, particularly during the winter season, are also projected across South Asia, and an increase in mean and extreme precipitation during the summer monsoon season is projected over India. Additionally, extreme hot days and nights are projected to increase across South Asia. Increased precipitation will be more pronounced over northern parts of South Asia, including Bangladesh and Sri Lanka, along with a slight decrease over Pakistan (Turner and Annamalai 2012). Finally, in the case of Southeast Asia, due to substantial variations in climate patterns caused by land-sea contrast and complex terrain at the sub-regional scale, the model simulations show substantial variations in predictions, such as a moderate increase in rainfall over the Indonesian islands (Christensen et al. 2013).

> **Box 1.5** *"I say 'welcome to climate change' when people come here,"* **said President Loek of Marshall Islands**
> Any discussion about the long term impacts of climate change in the future is incomplete without any discussion about rising sea levels, particularly for the low lying islands in the Pacific. The results from all model projections indicate varying levels of higher sea levels by the end of the twenty-first century as shown in the figure below. However, some of these islands may become uninhabitable long before they actually get submerged, due to saline contamination and ruining of crops. For instance Tuvalu located in the tropics, which is predominantly made of corals with an average elevation of 2 m. It consists of nine low lying islands, which are are already experiencing the impacts of sea level rise. Each year they experience "King Tide" when flood water comes

up out of the ground and destroys local crops, fresh water, and damages properties. Most residents in Tuvalu don't have flood insurance to protect their properties from flood damage.

A recent study by Becker et al. (2011) examined sea level rise in the western Pacific Ocean using a combination of tidal gauges, satellite based measurements, ocean modelling and GPS. The study showed that this region is experiencing sea level rise much larger than the global average. Specifically, in the case of Funafati island the total sea level rise from 1950 to 2009 was 5.1 ($\pm$0.7) mm per year, which is almost 3 times greater than the global average for the same period.

Similarly, the Marshall Islands, a group of 29 atolls and coral islands with an average elevation of 2 m above sea level, had to declare a state of emergency in 2013 when they were faced with droughts and some of the worst floods in their history. Approximately, 6000 people had to live on one liter of water per day. The physical disappearance of these island nations is a real threat that is being debated in international forums concerning options to protect the citizens or relocate them to nations in neighboring countries such as Australia and New Zealand. The Maldives, the world's lowest lying island chain, are already facing the effects of saltwater encroachment through the contamination of groundwater. The Maldives have already started preparing for sea level rise by building coastal infrastructure such as the flood-proofing of waste management systems and the protection of critical seaside vegetation.

Conclusions

Based on the discussion above it is evident that all of the Global South is in extremely vulnerable to already occurring and impending impacts of climate change. Some areas are more vulnerable than others, such as sea level rise in the low lying islands in the Pacific, extended periods of droughts in Northern and Sub Saharan Africa due to decline in precipitation, and extreme weather events like heat waves, floods, and hurricanes. In addition, most of the Global South is not adequately prepared for such changes resulting from climate change in the near future. This is mainly due to its high population density in more vulnerable areas, and limited access to resources and poor infrastructure for adequate coping strategies. Recently, several major storms such as Typhoon Hainan and Nargis, and Cyclone Pam have exposed the amount of destruction and losses caused to humanity, and the amount of time it takes to recover from their impacts. The effects of climate change are also being experienced indirectly in the form of the spread of infectious diseases in the highlands of Ethiopia, or failure of crop yields due to untimely rainfall in parts of Asia.

In this context, it is important to acknowledge that the impacts of climate change are not gender neutral. The negative effects of long term climate change in the Global South are disproportionately greater on women than men. For instance, about 70% of the world's poor are women, most of whom live in areas extremely vulnerable to climate change in the Global South. A major proportion of responsibilities relating to household water collection, food production, and preparation are on women. Thus they are exposed to the vagaries of climate extremes, which in some cases also results in limited access to financial resources and livelihood strategies. As a result, it is important to involve women in decision-making processes at all levels and highlight the impacts of long-term trends in climate change related processes in the Global South.

In view of the importance of developing more gender sensitive policies, during the seventh and eighteenth session of the Conference of Parties (COP), the decision of promoting gender equality and promoting participation of women in UNFCCC (United Nation Framework Convention on Climate Change) negotiations was adopted. Since 2012, Gender Day has been observed on December 9th during the COP sessions of UNFCCC in collaboration with governments, United Nations entities, intergovernmental organizations, and civil societies to promote gender sensitive policies for climate change. For instance in the recently concluded COP 20 session in Lima, Peru, a high level meeting among government representatives was held on gender and climate change. Some of the specific goals during that meeting included calls for action toward achieving gender equality. The meeting also called for enhancing awareness about the Beijing Declaration and Platform for Action regarding gender equality and climate change. However, there is a critical need to assess the regional to local level variations in gender gaps and inequalities vis a vis the impacts of climate change at various spatial scales. Therefore, in the next chapter an in-depth analysis of gender gaps and inequalities across the Global South has been conducted in order to facilitate a more gender sensitive understanding of the impacts of climate change.

References

Achard, F., Eva, H. D., Stibig, H. J., Mayaux, P., Gallego, J., Richards, T., et al. (2002). Determination of deforestation rates of the world's humid tropical forests. *Science, 297*(5583), 999–1002.

Barros, V. R., Doyle, M., & Camilloni, I. (2008). Precipitation trends in southeastern South America: Relationship with ENSO phases and the low-level circulation. *Theoretical and Applied Climatology, 93,* 19–33.

Baumgardner, D., Raga, G. B., & Muhlia, A. (2004). Evidence for the formation of CCN by photochemical processes in Mexico City. *Atmospheric Environment, 38,* 357–367.

Becker, A., Finger, P., Meyer-Christoffer, A., Rudolf, B., & Ziese, M. (2011). *GPCC full data reanalysis version 7.0 at 0.5: Monthly land-surface precipitation from raingauges built on GTS-based and historic data.*

Biasutti, M., & Sobel, A. H. (2009). Delayed seasonal cycle and African monsoon in a warmer climate. *Geophysical Research Letters, 36,* L23707.

Bonan, G. B. (2008). Forests and climate change: Forcings, feedbacks, and the climate benefits of forests. *Science, 320*(5882), 1444–1449.

Bousquet, P. (2011). Source attribution of the changes in atmospheric methane for 2006–2008. *Atmospheric Chemistry and Physics, 10,* 27603–27630.

Bracegirdle, T. J., Shuckburgh, E., Sallee, J.-B., Wang, Z., Meijers, A. J. S., Bruneau, N., et al. (2013). Assessment of surface winds over the Atlantic, Indian, and Pacific Ocean sectors of

References

the Southern Ocean in CMIP5 models: historical bias, forcing response, and state dependence. *Journal of Geophysical Research: Atmospheres, 118*(2), 547–562.

Carslaw, K. S., Boucher, O., Spracklen, D. V., Mann, G. W., Rae, J. G. L., Woodward, S., et al. (2010). A review of natural aerosol interactions and feedbacks within the Earth system. *Atmospheric Chemistry and Physics, 10*(4), 1701–1737.

Chagnon, F. J. F., & Bras, R. L. (2005). Contemporary climate change in the Amazon. *Geophysical Research Letters, 32*, L13703. https://doi.org/10.1029/2005GL022722.

Choi, G., Collins, D., Ren, G., Trewin, B., Baldi, M., Fukuda, Y., et al. (2009). Changes in means and extreme events of temperature and precipitation in the Asia-Pacific Network region, 1955–2007. *International Journal of Climatology, 29*(13), 1906–1925.

Christensen, J. H., Kanikicharla, K. K., Marshall, G., & Turner, J. (2013). Climate phenomena and their relevance for future regional climate change. In T. F. Stocker, D. Qin, G.-K. Plattner, M. Tignor, S. K. Allen, J. Boschung, et al. (Eds.), *Climate change 2013: The physical science basis contribution of working group I to the Fifth Assessment Report of the Intergovernmental Panel on Climate Change* (pp. 1217–1308). Cambridge: Cambridge University Press.

Comarazamy, D. E., & Gonzalez, J. E. (2011). Regional long-term climate change (1950–2000) in the midtropical Atlantic and its impacts on the hydrological cycle of Puerto Rico. *Journal of Geophysical Research Atmospheres, 116*. https://doi.org/10.1029/2010jd015414.

Conway, D., Hanson, C. E., Doherty, R., & Persechino, A. (2007). GCM simulations of the Indian Ocean dipole influence on East African rainfall: Present and future. *Geophysical Research Letters, 34*(3), L03705.

de Miranda, R. M., Andrade, M. D., Worobiec, A., & Van Grieken, R. (2002). Characterisation of aerosol particles in the Sao Paulo Metropolitan Area. *Atmospheric Environment, 36*, 345–352.

Defries, R. S., Bounoua, L., & James Collatz, G. (2002). Human modification of the landscape and surface climate in the next fifty years. *Global Change Biology, 8*(5), 438–458.

Deser, C., Phillips, A., Bourdette, V., & Teng, H. (2012). Uncertainty in climate change projections: the role of internal variability. *Climate Dynamics, 38*(3-4), 527–546.

Dlugokencky, E., et al. (2009). Observational constraints on recent increases in the atmospheric CH4 burden. *Geophysical Research Letters, 36*, L18803.

Donat, M. G., Alexander, L. V., Yang, H., Durre, I., Vose, R., Dunn, R. J. H., et al. (2013a). Updated analyses of temperature and precipitation extreme indices since the beginning of the twentieth century: the HadEX2 dataset. *Journal of Geophysical Research: Atmospheres, 118*(5), 2098–2118.

Donat, M. G., Alexander, L. V., Yang, H., Durre, I., Vose, R., & Caesar, J. (2013b). Global land-based datasets for monitoring climatic extremes. *Bulletin of the American Meteorological Society, 94*(7), 997–1006.

Douglas, E. M., Wood, S., Sebastian, K., Vörösmarty, C. J., Chomitz, K. M., & Tomich, T. P. (2007). Policy implications of a pan-tropic assessment of the simultaneous hydrological and biodiversity impacts of deforestation. *Water Resources Management, 21*(1), 211–232.

Erel, Y. G., Dayan, U., Rabi, R., Rudich, Y., & Stein, M. (2006). Trans boundary transport of pollutants by atmospheric mineral dust. *Environmental Science & Technology, 40*, 2996–3005.

Exbrayat, J.-F., & Williams, M. (2015). Quantifying the net contribution of the historical Amazonian deforestation to climate change. *Geophysical Research Letters, 42*(8), 2968–2976.

FAO. (2010). Global forest resources assessment 2010 main report. *FAO Forestry Paper 163*.

Fyfe, J. C. (2003). Separating extratropical zonal wind variability and mean change. *Journal of Climate, 16*(5), 863–874.

Ghose, M. K., Paul, R., & Banerjee, R. K. (2005). Assessment of the status of Urban air pollution and its impact on human health in the city of Kolkata. *Environmental Monitoring and Assessment, 108*, 151–167.

Gillett, N. P., Kell, T. D., & Jones, P. D. (2006). Regional climate impacts of the Southern Annular Mode. *Geophysical Research Letters, 33*(23), L23704.

Giorgi, F., Im, E.-S., Coppola, E., Diffenbaugh, N. S., Gao, X. J., Mariotti, L., et al. (2011). Higher hydroclimatic intensity with global warming. *Journal of Climate, 24*(20), 5309–5324.

Goswami, B. N., Venugopal, V., Sengupta, D., Madhusoodanan, M. S., & Xavier, P. K. (2006). Increasing trend of extreme rain events over India in a warming environment. *Science, 314*(5804), 1442–1445.

Hartmann, D. L., Klein Tank, A. M. G., Rusticucci, M., Alexander, L. V., Brönnimann, S., Charabi, Y., et al. (2013). Observations: Atmosphere and surface. In T. F. Stocker, D. Qin, G.-K. Plattner, M. Tignor, S. K. Allen, J. Boschung, et al. (Eds.), *Climate change 2013: The physical science basis contribution of working group I to the Fifth Assessment Report of the Intergovernmental Panel on Climate Change.* Cambridge: Cambridge University Press.

Jhajharia, D., Shrivastava, S., Sarkar, D., & Sarkar, S. (2009). Temporal characteristics of pan evaporation trends under humid conditions of northeast India. *Agriculture Forest Meteorology, 336*, 61–73.

Jiang, Z., Song, J., Li, L., Chen, W., Wang, Z., & Wang, J. (2012). Extreme climate events in China: IPCC-AR4 model evaluation and projection. *Climatic Change, 110*(1-2), 385–401.

Jonsson, P. (2004). Vegetation as an urban climate control in the subtropical city of Gaborone, Botswana. *International Journal of Climatology, 24*, 1307–1322.

Juma, M. (2010). Security and regional cooperation in Africa: how can we make Africa's security architecture fit for the new challenges. In *Climate change resources migration: Securing Africa in an uncertain climate.* Cape Town: Heinrich Böll Foundation Southern Africa.

Jung, M., Reichstein, M., Ciais, P., Seneviratne, S. I., Sheffield, J., Goulden, M. L., et al. (2010). Recent decline in the global land evapotranspiration trend due to limited moisture supply. *Nature, 467*(7318), 951–954.

Karl, T. R., Arguez, A., Huang, B., Lawrimore, J. H., McMahon, J. R., Menne, M. J., et al. (2015). Possible artifacts of data biases in the recent global surface warming hiatus. *Science, 348*, 1469.

Kaufmann, R. K., Kauppi, H., Mann, M. L., & Stock, J. H. (2011). Reconciling anthropogenic climate change with observed temperature 1998–2008. *Proceedings of the National Academy of Sciences of the United States of America, 108*, 11790–11793.

Koomey, J. (2015). Why we can't ditch the 2 C warming goal. *Ecowatch Blog.* Retrieved July 10, 2015, from http://ecowatch.com/2014/10/27/2-c-warming-limit-climate-change/2/

Kosaka, Y., & Xie, S.-P. (2013). Recent global-warming hiatus tied to equatorial Pacific surface cooling. *Nature, 501*(7467), 403.

Krishna Moorthy, K., Suresh Babu, S., Manoj, M. R., & Satheesh, S. K. (2013). Buildup of aerosols over the Indian region. *Geophysical Research Letters, 40*(5), 1011–1014.

Krishnan, R., & Sugi, M. (2003). Pacific decadal oscillation and variability of the Indian summer monsoon rainfall. *Climate Dynamics, 21*(3-4), 233–242.

Kueppers, L. M., Snyder, M. A., Sloan, L. C., Cayan, D., Jin, J., Kanamaru, H., et al. (2008). Seasonal temperature responses to land use change in the western United States. *Global Planetary Change, 60*, 250–264.

Kumar, K. K., Rajagopalan, B., Hoerling, M., Bates, G., & Cane, M. (2006). Unraveling the mystery of Indian monsoon failure during El Niño. *Science, 314*(5796), 115–119.

Lee, T.-c., Chan, K.-y., Chan, H.-s., & Kok, M.-h. (2011). Projections of extreme rainfall in Hong Kong in the 21st century. *Acta Meteorologica Sinica, 25*, 691–709.

Lejeune, Q., Davin, E. L., Guillod, B. P., & Seneviratne, S. I. (2015). Influence of Amazonian deforestation on the future evolution of regional surface fluxes, circulation, surface temperature and precipitation. *Climate Dynamics, 44*(9-10), 2769–2786.

Li, C. S., & Lin, C. H. (2002). PM1/PM2.5/PM10 characteristics in the urban atmosphere of Taipei. *Aerosol Science and Technology, 36*, 469–473.

Lindén, J. (2011). Nocturnal Cool Island in the Sahelian city of Ouagadougou, Burkina Faso. *International Journal of Climatology, 31*, 605–620. https://doi.org/10.1002/joc.2069.

Lobell, D. B., Bonfils, C. J., Kueppers, L. M., & Snyder, M. A. (2008). Irrigation cooling effect on temperature and heat index extremes. *Geophysical Research Letters, 35*, L09705. https://doi.org/10.1029/2008GL034145.

Lobell, D., Bala, G., Mirin, A., Phillips, T., Maxwell, R., & Rotman, D. (2009). Regional differences in the influence of irrigation on climate. *Journal of Climate, 22*(8), 2248–2255.

References

Mahmood, R., Pielke, R. A., Hubbard, K. G., Niyogi, D., Dirmeyer, P. A., McAlpine, C., et al. (2014). Land cover changes and their biogeophysical effects on climate. *International Journal of Climatology, 34*(4), 929–953.

Marengo, J. A., Liebmann, B., Grimm, A. M., Misra, V., Silva Dias, P. L., Cavalcanti, I. F. A., et al. (2012). Recent developments on the South American monsoon system. *International Journal of Climatology, 32*(1), 1–21.

Marengo, J. A. (2004). Interdecadal variability and trends of rainfall across the Amazon basin. *Theoretical and Applied Climatology, 78*(1–3), 79–96.

May, W. (2011). The sensitivity of the Indian summer monsoon to a global warming of 2 C with respect to pre-industrial times. *Climate Dynamics, 37*(9-10), 1843–1868.

Mazzeo, N. A., & Venega, L. E. (2004). Some aspects of air pollution in Buenos Aires city. *International Journal of Environment and Pollution, 22*, 365–378.

Meehl, G. A., Covey, C., Delworth, T., Latif, M., McAvaney, B., Mitchell, J., et al. (2007). The WCRP CMIP3 multi-model dataset: A new era in climate change research. *Bulletin of the American Meteorological Society, 88*, 1383–1394.

Meehl, G. A., & Arblaster, J. M. (1998). The Asian-Australian monsoon and el niño-southern oscillation in the NCAR climate system model*. *Journal of Climate, 11*(6), 1356–1385.

Meehl, G. A., et al. (2011). Model-based evidence of deep-ocean heat uptake during surface-temperature hiatus periods. *Nature Climate Change 1.7*, 360.

Meehl, G. A., Hu, A., Arblaster, J., Fasullo, J., & Trenberth, K. E. (2013). Externally forced and internally generated decadal climate variability associated with the Interdecadal Pacific oscillation. *Journal of Climate*. https://doi.org/10.1175/JCLI-D-12-00548.1.

Mishra, V., Ganguly, A. R., Nijssen, B., & Lettenmaier, D. P. (2015). Changes in observed climate extremes in global urban areas. *Environmental Research Letters, 10*(2), 024005.

Moss, R. H., Edmonds, J. A., Hibbard, K. A., Manning, M. R., Rose, S. K., Van Vuuren, D. P., et al. (2010). The next generation of scenarios for climate change research and assessment. *Nature, 463*(7282), 747–756.

Neelin, J. D., Münnich, M., Hui, S., Meyerson, J. E., & Holloway, C. E. (2006). Tropical drying trends in global warming models and observations. *Proceedings of the National Academy of Sciences, 103*(16), 6110–6115.

Nicholson, S. E., & Kim, J. (1997). The relationship of the El Niño–Southern Oscillation to African rainfall. *International Journal of Climatology, 17*, 117–135.

O'Shea, P. M., Roy, S. S., & Singh, R. B. (2015). Diurnal variations in the spatial patterns of air pollution across Delhi. *Theoretical and Applied Climatology, 124*, 609. https://doi.org/10.1007/s00704-015-1441-y.

Piao, S., Ciais, P., Huang, Y., Shen, Z., Peng, S., Li, J., Zhou, L., et al. (2010). The impacts of climate change on water resources and agriculture in China. *Nature, 467*(7311), 43–51.

Population Reference Bureau. (2015). *World population data sheet 2014*. Retrieved June 24, 2015, from http://www.prb.org/pdf14/2014-world-population-data-sheet_eng.pdf

Quintana, J. M., & Aceituno, P. (2012). Changes in the rainfall regime along the extratropical west coast of South America (Chile): 30–43oS. *Atmosfera, 25*, 1–22.

Quintana-Gomez, R. A. (1999). Trends of maximum and minimum temperatures in northern South America. *Journal of Climate, 12*(7), 2104–2112.

Ren, G. Y., Chu, Z. Y., Chen, Z. H., & Ren, Y. Y. (2007). Implications of temporal change in urban heat island intensity observed at Beijing and Wuhan stations. *Geophysical Research Letters, 34*, L05711.

Rigby, M., et al. (2008). Renewed growth of atmospheric methane. *Geophysical Research Letters, 35*, L22805.

Roderick, M. L., Rotstayn, L. D., Farquhar, G. D., & Hobbins, M. T. (2007). On the attribution of changing pan evaporation. *Geophysical Research Letters, 34*, L17403.

Rohde, R., Muller, R. A., Jacobsen, R., Muller, E., Perlmutter, S., Rosenfeld, A., et al. (2013). A new estimate of the average Earth surface land temperature spanning 1753 to 2011. *Geoinformation Geostatistics: An Overview, 1*. https://doi.org/10.4172/gigs.1000101.

24 1 Climate Change in the Global South: Trends and Spatial Patterns

Schnadt Poberaj, C., Staehelin, J., Brunner, D., Thouret, V., De Backer, H., & Stübi, R. (2009). Long-term changes in UT/LS ozone between the late 1970s and the 1990s deduced from the GASP and MOZAIC aircraft programs and from ozonesondes. *Atmospheric Chemistry and Physics, 9*(14), 5343–5369.

Sen Roy, S., & Balling, R. C. (2005). Analysis of trends in maximum and minimum temperature, diurnal temperature range, and cloud cover over India. *Geophysical Research Letters, 32*, L12702.

Sen Roy, S., Mahmood, R., Niyogi, D., Lei, M., Foster, S. A., Hubbard, K. G., et al. (2007). Impacts of the agricultural Green Revolution–induced land use changes on air temperatures in India. *Journal of Geophysical Research: Atmospheres, 112*, D21.

Sen Roy, S., Singh, R. B., & Kumar, M. (2011). An analysis of local-spatial temperatures patterns in the Delhi metropolitan area. *Physical Geography, 32*, 114–138.

Sen Roy, S. (2006). The impacts of ENSO, PDO, and local SSTs on winter precipitation in India. *Physical Geography, 27*(5), 464–474.

Sen Roy, S. (2009). A spatial analysis of extreme hourly precipitation patterns in India. *International Journal of Climatology, 29*(3), 345–355.

Sen Roy, S., & Rouault, M. (2013). Spatial patterns of seasonal scale trends in extreme hourly precipitation in South Africa. *Applied Geography, 39*, 151–157.

Seneviratne, S. I., Nicholls, N., Easterling, D., Goodess, C. M., Kanae, S., Kossin, J., et al. (2012). Changes in climate extremes and their impacts on the natural physical environment. In C. B. Field et al. (Eds.), *Managing the risks of extreme events and disasters to advance climate change adaptation: A special report of working groups I and II of the Intergovernmental Panel on Climate Change (IPCC)* (pp. 109–230). Cambridge: Cambridge University Press.

Seth, A., Rojas, M., & Rauscher, S. A. (2010). CMIP3 projected changes in the annual cycle of the South American Monsoon. *Climatic Change, 98*(3-4), 331–357.

Sham, S. (1987). The urban heat island – its concept and application to Kuala Lumpur. In S. Sham (Ed.), *Urbanisation and the atmospheric environment in the low tropics: Experiences from the Kelang Valley Region, Malaysia.* Malaysia: Pernerbit University Kebangsaan.

Solomon, S., Daniel, J. S., Neely, R. R., Vernier, J. P., Dutton, E. G., & Thomason, L. W. (2011). The persistently variable "background" stratospheric aerosol layer and global climate change. *Science, 333*, 866–870.

Solomon, S., Rosenlof, K. H., Portmann, R. W., Daniel, J. S., Davis, S. M., Sanford, T. J., et al. (2010). Contributions of stratospheric water vapor to decadal changes in the rate of global warming. *Science, 327*, 1219–1223.

Stern, N. (2007). *The economics of climate change: The Stern review.* Cambridge: Cambridge University Press.

Stocker, T. F., Qin, D., Plattner, G.-K., Alexander, L. V., Allen, S. K., Bindoff, N. L., et al. (2013). Technical summary. In T. F. Stocker, D. Qin, G.-K. Plattner, M. Tignor, S. K. Allen, J. Boschung, A. Nauels, Y. Xia, V. Bex, & P. M. Midgley (Eds.), *Climate change 2013: The physical science basis contribution of working group I to the Fifth Assessment Report of the Intergovernmental Panel on Climate Change.* Cambridge: Cambridge University Press.

Tanner, P. A., & Law, P. T. (2003). Organic acids in the atmosphere and bulk deposition of Hong Kong. *Water Air and Soil Pollution, 142*, 279–297.

Tans, P. (2009). An accounting of the observed increase in oceanic and atmospheric CO2 and an outlook for the future. *Oceanography, 22*, 26–35.

Thompson, L. G., et al. (2009). Glacier loss on Kilimanjaro continues unabated. *Proceedings of the National Academy of Sciences* 106.47: 19770–19775.

Trenberth, K. E., Jones, P. D., Ambenje, P., Bojariu, R., Easterling, D., Klein Tank, A., et al. (2007). Observations: Surface and atmospheric climate change. In S. Solomon, D. Qin, M. Manning, Z. Chen, M. Marquis, K. B. Averyt, M. Tignor, & H. L. Miller (Eds.), *Climate change 2007: The physical science basis. Contribution of working group I to the Fourth Assessment Report of the Intergovernmental Panel on Climate Change.* Cambridge: Cambridge University Press.

Trenberth, K. E., & Fasullo, J. T. (2013). An apparent hiatus in global warming? *Earth's Future, 1*(1), 19–32.

Turner, A. G., & Annamalai, H. (2012). Climate change and the South Asian summer monsoon. *Nature Climate Change, 2*(8), 587–595.

United Nations. (2016). SDGs: Sustainable development knowledge platform. *United Nations*. Retrieved December 29, 2016, from https://sustainabledevelopment.un.org/sdgs

Villar, J. C. E., Guyot, J. L., Ronchail, J., Cochonneau, G., Filizola, N., Fraizy, P., et al. (2009). Contrasting regional discharge evolutions in the Amazon basin (1974–2004). *Journal of Hydrology, 375*(3), 297–311.

Vose, R. S., Easterling, D. R., & Gleason, B. (2005). Maximum and minimum temperature trends for the globe: An update through 2004. *Geophysical Research Letters, 32*, L23822.

WHO. (2015). *7 million premature deaths annually linked to air pollution*. Retrieved June 26, 2015, from http://www.who.int/mediacentre/news/releases/2014/air-pollution/en/

Wild, M., Grieser, J., & Schaer, C. (2008). Combined surface solar brightening and increasing greenhouse effect support recent intensification of the global land-based hydrological cycle. *Geophysical Research Letters, 35*, L17706.

Williams, C. A., & Hanan, N. P. (2011). ENSO and IOD teleconnections for African ecosystems: evidence of destructive interference between climate oscillations. *Biogeosciences, 8*(1), 27–40.

World Bank. (2015). *Little data book on climate change: Supplemental data*. Retrieved June 29, 2015, from http://data.worldbank.org/data-catalog/ldbcc-supplemental

Yan, Z., Li, Z., Li, Q., & Jones, P. (2010). Effects of site change and urbanisation in the Beijing temperature series 1977–2006. *International Journal of Climatology, 30*, 1226–1234.

Zhang, Y., Liu, C., Tang, Y., & Yang, Y. (2007). Trends in pan evaporation and reference and actual evapotranspiration across the Tibetan Plateau. *Journal of Geophysical Research: Atmospheres, 112*(D12).

Chapter 2
Spatial Patterns of Gender Inequalities/ Inequities Across the Global South

> Gender equality is more than a goal in itself. It is a precondition for meeting the challenge of reducing poverty, promoting sustainable development and building good governance.
> —Kofi Annan

Development Goal 5: "Achieve gender equality and empower all women and girls" (UN 2015).

Introduction

On December 10, 1948 the Universal Declaration of Human Rights was adopted which established that all people regardless of their gender are equal in dignity and rights. This was followed by agreements, with one of the most notable of those agreements being the UN Convention on the Elimination of All Forms of Discrimination against Women (CEDAW) adopted in 1979 by the UN General Assembly. This agreement was created specifically to guarantee women and girls equal rights. By ratifying the CEDAW agreement, 186 nations committed to getting rid of any form of discrimination against women. This would ensure equality among men and women, and ensuring equal access to, and equal opportunities in the political, social, economic, civil or any other field. It endorsed empowerment of women and their full participation, including decision making processes and access to power. However, a recent analysis by UCLA's World Policy Analysis Center revealed that the constitutions in 6% of the nations who adopted CEDAW allow customary or religious law to supersede constitutional provisions, thus undermining the constitution and protection of women and girls' rights. Additionally, only 75% of the CEDAW countries provide secondary education through completion for girls and women.

The next major agreement among nations was the Beijing declaration in 1995, where 189 nations reaffirmed their commitments to equal rights and human dignity of women and men, and ensured the full implementation of human rights and fundamental freedoms. In a recent meeting held in March 2015, titled Beijing+20, the Executive Director of UN Women, Ms. Phumzile Mlambo-Ngcuka, highlighted the slow and patchy progress toward gender equality over the last 20 years. She specifically highlighted the drop in maternal mortality, as well as an increase in girls' enrollment in schools. However, there still remain gaping gender inequalities, some of which are listed below:

- Less than 20% of the world's landholders are women. Women represent fewer than 5% of all agricultural landholders in North Africa and West Asia, while in sub-Saharan Africa they make up an average of 15% (UN Women 2015).
- Women spend 16 million h a day collecting drinking water in 25 sub-Saharan countries, compared to 6 million h a day for men, and 4 million h a day for children (WHO and UN Children's Fund 2015).
- There was a 45% decrease in maternal deaths compared to 1990. However, 800 women still die every day from preventable pregnancy-related causes. Among these, 99% occur in developing countries located in the Global South (UNFPA 2015).
- Despite the increase in girls in schools, the gender disparity widens at the secondary and tertiary levels as well as in STEM disciplines in higher education in many countries. For instance, in Sub Saharan Africa there are 64 girls per 100 boys at the tertiary level of education.
- Additionally, 60% of the world's illiterate are women who are mostly concentrated in the developing and less developed countries in the Global South (UN Women 2015).
- In 2000 the UN Security Council passed resolution 1325, which recognized that the impacts of war are not gender neutral, rather that women are affected more adversely. Therefore, through this resolution greater participation of women was stressed. However, in reality between 1992 and 2011 only 9% of negotiators at the peace tables were women.
- Despite the 1993 UN General Assembly declaration on the elimination of violence against women, more than 1 in 3 women experience physical or sexual violence.
- Four million women die from unsafe cooking conditions every year (Clinton Foundation 2015).
- Finally, despite the increased presence of women as news subjects in print, radio, and television, only 6% of the stories challenge gender stereotypes compared to 46% of the stories which continue to support and even reinforce those stereotypes (UNESCO 2015).

In this context, it is noteworthy that one of the eight principal goals for 2015 chartered by the Millennium Development Goals is to promote gender equality and empower women. Thus there is greater emphasis on gender mainstreaming methodology, which consists of integrating a gendered approach into development and environmental efforts (UNCCD 2015). It includes women's participation in existing strategies and programs. Furthermore, there is greater emphasis on gender-related issues in climate change dialogues, programs, and projects (Agostine and Lizarde

2012; Brody et al. 2008; Dankelman 2010; Denton 2002; Terry 2009). The disproportionate burden of long-term impacts of climate change on women has been recently highlighted in reports published by various UN agencies, including United Nations Development Program (UNDP), Food and Agricultural Organization (FAO), World Health Organization (WHO), and United Nations Educational, Scientific and Cultural Organization (UNESCO). These reports focus on specific impacts of climate change on women in various regions of the Global South highlighting issues such as human health, agriculture and food security, and indoor pollution. They further highlight the importance of incorporating the differential impacts of long-term climate change on women, and thereby encourage the participation of women in decision-making processes at the local level. More importantly, this gender sensitive approach is particularly significant in South Asia, which is undergoing a massive transformation in its socio-ecological sphere related to climate change, while traditional patriarchal societies face strong challenges to their gendered power structures (Sultana 2014).

> **Box 2.1 UN Women**
>
>
>
> The UN Women, started in 2011, is part of United Nations. This organization is specifically dedicated to gender equality and empowerment of women and girls. Some of the main objectives of this organization include elimination of discrimination of women and achievement of equality between women and men as partners and beneficiaries of development. It also focuses on equal human rights, humanitarian action, peace, and security. The main goals of UN Women as stated on their website are listed below:
>
> - To support inter-governmental bodies, such as the Commission on the Status of Women, in their formulation of policies, global standards and norms.
> - To help Member States to implement these standards, standing ready to provide suitable technical and financial support to those countries that request it, and to forge effective partnerships with civil society.
> - To lead and coordinate the UN system's work on gender equality as well as promote accountability, including through regular monitoring of system-wide progress.

Variability in climatic conditions impacts human and physical systems at different geographic scales, and is further complicated by local environmental factors and topography, which affect the vulnerability of resident populations. Negative impacts of climate change are also experienced: within relatively small time periods in the form of natural hazards—landslides, floods, hurricanes; and in the longer run by the gradual degradation of the immediate environment—damaging effects of long-term increases in temperatures and the frequency of extreme weather events. Adverse impacts of climate change are already evident on agricultural productivity and food security in response to worsening droughts, increasingly vulnerable biodiversity and ecosystems, and stressed water resources. Additionally, changing climatic conditions over different time scales are impacting human health in the form of infectious diseases spreading across newer territories and settlements, migration patterns, energy, transport, and industry. These detrimental impacts of climate change, especially in developing countries, often affect women more than men. A clear majority of the world's 2.2 billion people, approximately 20% of the world's total population, are living in or near multidimensional poverty (UNDP 2015a, b, c, d). Most of those living under poverty are women, who rely most heavily on those natural resources susceptible to climate change (UN Population Fund 2015). In many cases, women also face greater social, economic, and political challenges than males, which limit their coping capacity. Some of the specific issues can be summarized as follows:

- In rural areas women are usually responsible for collecting water for the families as well as fuelwood for cooking and heating. Specifically, in almost two-thirds of the countries, women and girls are primarily responsible for obtaining water for their households and communities (UNDP 2015a, b, c, d). They often have to walk longer distances to collect water and fuelwood due to lengthening and intensified droughts resulting from deforestation and resource depletion. As distances to water sources lengthen due to increasing droughts, women become more vulnerable to violence and to the vagaries of harsh weather conditions, such as high temperatures. Not surprisingly, this detracts from the time they can spend getting and completing their education.
- Unequal access to resources and decision-making processes, along with limited mobility, exposes women in rural areas to more of the negative effects of climate change. Often, in view of the traditionally limited role of women in decision-making processes at the household, village, and national levels in most cultures, their needs, interests, and constraints are poorly reflected in policy-making processes and government programs aimed at poverty reduction, food security, and environmental sustainability (FAO 2015).
- Mortality associated with indoor smoke from burning solid fuel amounts to nearly 4.3 million deaths per year, with women and children in rural areas at greatest risk (WHO 2015). About 25% of these fatalities occur in India, where women and children spend more time in homes that burn fuel wood and coal for cooking and heating purposes. Furthermore, it has been noted that some 80% of rural women in Asia are affected by the growing shortage of firewood (UNDP 2015a, b, c, d).
- A greater proportion of women in poor countries engage in subsistence farming and water collection, exposing them more adversely to the effects of environmental degradation in the form of food shortages and malnutrition (UNDP 2015a, b, c, d).

Thus it is evident that gender differences in various aspects of social and economic settings have been highlighted widely in many published studies and popular media. Some of the commonly used terms to refer to these gendered differences include Gender Gap, Gender Inequity, and Gender Inequality. Gender Gap is usually used to refer to differences in voting behavior, income levels, empowerment, and other attitudes. Gender Inequality refers to unequal treatment of people based on gender, which is usually a result of social and cultural norms. The last term, Gender Inequity is close in definition to the two previous terms, and more specifically refers to unfair situations leading to gender inequality, gender based bias, or lack of equity. The key definition for Gender Inequity is unfair treatment based on gender. Additionally, in the recent past there has been awareness about gendered differences in the impacts of climate change and thus a push toward developing more gender sensitive policies. In order to gain a deeper understanding of the impacts of climate change and climate variability, the objective of this chapter is analyze the spatial variations of gender disparities across the Global South. A comparative study of the various indicators at the regional scale will enable us to identify the hotspots of steep gender disparities. In view of the emphasis on reducing gender disparities across the globe, there is a large amount of data, which in some cases are gender disaggregated, collected by leading organizations including the World Bank, Population Reference Bureau, FAO, UNDP, WHO, World Economic Forum, and others. In the following sections the spatial patterns of some of the major indicators of gender disparities are discussed.

Human Development Index (HDI)

Before getting into the specific gender-related indices, it is important to examine the regional scale trends in some of the general indices that reflect levels of development among various nations. In this context, one of the most widely used indices for doing an effective comparative analysis of development among nations at the global scale is the Human Development Index (HDI), which is published every year by the UNDP (UNDP 2015a, b, c, d). This index was created to highlight people and their capabilities as the criteria for assessing levels of development of different nations. It is also a good indicator of the effectiveness of various national level policies in promoting human development outcomes for different nations with the same economic levels. The HDI is designed to reflect average achievements in three basic aspects of human development—leading a long and healthy life, being knowledgeable, and enjoying a decent standard of living. Figure 2.1 shows the regional level change in HDI rankings from 1980 to 2013. In general all the regions experienced an overall improvement in the HDI rankings since it was first developed in 1980. Two regions, Central Asia and Europe and Latin America and the Caribbean, persistently showed the highest levels of HDI on the global scale. However, the lowest levels of HDI were observed over Sub Saharan Africa followed by South Asia. Globalization is considered to have produced major gains in the HDI worldwide, including advances

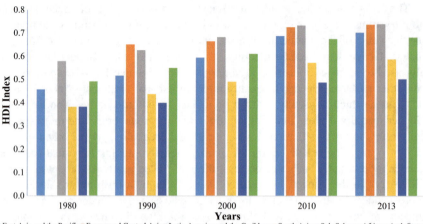

Fig. 2.1 Regional level changes in HDI Index 1980–2013

in technology, education, and income, which have led to better quality and standard of life, with greater security and healthier lives. The focus of the latest HDI 2014 report highlights the uncertainty of sustained levels of higher HDI in different countries due to natural processes, such as extreme weather, and economic slump, which can very quickly undermine sustained human development (UNDP 2015a, b, c, d). Other factors that can lead to instability in the HDI include national level policies, corruption and unresponsive state institutions, political threats, unstable societies characterized by theft, violence, and community tension, and neglect of public health. Therefore, it is not only important to achieve higher levels of human development but also critical to make them secure and accessible to all levels of society in the future. Furthermore, it is important to identify natural and human vulnerabilities, which are usually defined as exposure to risks for sustained long term increase in HDI ranks across nations. Overall certain sections of the population such as women, children, and elderly are more vulnerable in varying degrees everywhere in the world. This is further determined by the existing infrastructure and equitable access to resources.

In the context of vulnerabilities, the increasing uncertainty and impacts of already occurring and impending climate change processes is of critical importance. For instance, recent occurrences of severe hurricanes and storms in parts of Southeast Asia exposed the weaknesses of the existing infrastructure and the vulnerability of low income populations living in the low lying islands of Southeast Asia. In this regard, women are often more vulnerable as a result of patriarchal societies where women do not have equal rights, or equal access to resources and decision making. In addition, greater proportion of women live in poverty worldwide. Therefore, it is critical to have a gender differentiated approach for better, more effective policy formulation.

Gender Development Index (GDI) 33

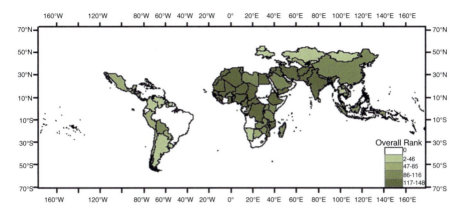

Fig. 2.2 Spatial patterns of GDI rankings across the Global South in 2014

Gender Development Index (GDI)

The GDI measures gender gap in human development achievements in three basic dimensions of human development, which include health, education, and command over economic resources. Health is measured by life expectancy at birth for males and females. The minimum and maximum years of life expectancy for females are 22.5 and 87.5 years, while for males it is 17.5 and 82.5 years. This is based on the average higher life expectancy of 5 years of females over males. Education is measured by expected years of schooling for children and mean years of schooling for adults aged 25 years and older for males and females. Finally, the command over economic resources is measured by the male and female estimated earned income (UNDP 2015a, b, c, d). The GDI is based on absolute deviation of gender parity from HDI, which implies equal ranking for gender gaps hurting males or females. In general the GDI ranking (indicative of higher gender gaps) was higher in countries with lower HDI ranking (UNDP 2015a, b, c, d).

The regional level spatial patterns of GDI across the Global South are depicted in Fig. 2.2. Higher GDI ranks are indicative of lower female HDI compared to the male HDI in those countries and vice versa. In general the lower rankings were clustered significantly across South America and parts of Africa. Only four countries in the Global South were ranked in the bottom 10 GDI rankings (indicative of lower gender inequalities), which included Argentina (2) and Venezuela (2) in South America, followed by Trinidad and Tobago (8) in the Caribbean, and Armenia (8) in Europe. There were no rakings available for some of the largest economies in the Global South, such as South Africa and Brazil. Most of the countries in South America were ranked were ranked in the top 50 for GDI, except Chile (61), Peru (72), and Bolivia (93). However, for most of Central America and the Caribbean, the GDI rankings were higher than 50, with two countries higher than 100, Nicaragua (102) and Guatemala (104). Across the African continent the lowest GDI rank was located in Namibia (36), followed by Lesotho (43). The rest of the countries across

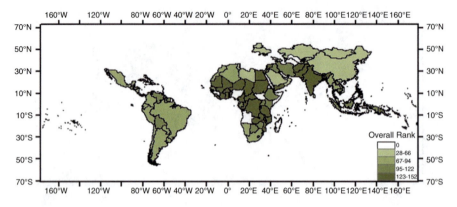

Fig. 2.3 Spatial patterns of GII rankings across the Global South in 2014

the continent were ranked higher than 50. The countries with the highest rank (indicative of high gender inequalities) across the African continent were mainly clustered in the Sahel region of Africa, including Niger (147), Chad (144), and Mali (143). These countries have predominantly semiarid climate regimes, which make them more vulnerable to the impacts of long term droughts and desertification. The impact of prolonged droughts are greater on women, mainly due to the stress caused by the scarcity of water and fuel wood. In general, all countries in Asia experienced higher GDI ranks, with an average value of 65. Only four countries were ranked within the lower 50 ranks of GDI (relatively lower gender inequalities), which include Thailand (14), Kazakhstan (25), Mongolia (32), and Kyrgyzstan (39). The highest ranking countries, indicative of the widest gaps in gender development, were clustered in South Asia, consisting of Afghanistan (148), Pakistan (145), and India (132). It is noteworthy that higher GDI ranking is indicative of unequal access to educational opportunities for women and children, quality of health care, and employment opportunities.

Gender Inequality Index (GII)

It is important to measure gender inequality, because it remains a major barrier to human development. Therefore countries with higher GII ranks usually have a lower HDI index. In many countries women and girls do not have equal access to health care, education, political representation, and access to decision to making processes, and in some cases equal rights. This index is built on the same framework as the HDI to better expose differences in the distribution of achievements between women and men. The objective of this index is to measure the human development costs of gender inequality (UNDP 2015a, b, c, d). A higher rank of GII is indicative of higher gender inequality. This index measures gender inequalities in three aspects of human development which include reproductive health, empowerment, and economic status. The reproductive health is measured by the maternal mortality ratio and adolescent birth rates. Empowerment consists of the proportion of female occupied parliamentary seats and proportion of adult females and males aged 25 years and older with some secondary education. Finally, the economic status consists of labor market participation, and is measured by labor force participation rate of females and male aged greater than or equal to 15 years (UNDP 2015a, b, c, d).

The spatial patterns of GII across the Global South are shown in Fig. 2.3. It has been widely emphasized that gender inequality is the leading factor hampering human development. GII is different from GDI, because GDI is aimed at assessing gender related development that is more suitable for more developed countries due to the indicators used in calculating the GDI. Some of the specific shortcomings of GDI include the high emphasis on the earned-income component, and the great neglect of gaps in education and life expectancy. Additionally, the earned-income has several shortcomings in terms of theoretical and practical aspects, which excessively penalizes the rich countries (Bardhan and Klasen 1999). As highlighted above the calculation of GII is related to the current gender achievement and distance from the baseline of equality. Thus a higher development index may not always be indicative of lower gender inequalities.

The average GII ranking across the Global South is 73, with lowest rankings observed in Belarus (28) followed by China (37), while the highest rankings were located in Yemen (152), Niger (151), and Chad (150). At the regional scale, in Central and South America, the lowest ranks were observed in the Bahamas (53) followed by Trinidad and Tobago (56). Some of the higher gender inequalities were also concentrated in the Caribbean and Central America: Haiti (132), Guyana (113), and Guatemala (112). The GII rank in most of the countries in South America was between 50 and 100, with the highest rank observed in Bolivia (97). Due to the constraints of the input factors, the GII was not calculated for several countries in Africa, including Angola, Guinea, Guinea Bissau, and Nigeria. Only two countries, Libya (40) and Tunisia (48), were ranked among the top 50 in the world, followed by 6 countries ranked among the top 100. The rest of the countries in Africa were ranked above 100. The highest rankings were observed in Niger (151) and Chad (150), all of which are located in the Sahel region of Africa. Overall, the higher ranks are mostly concentrated spatially in the middle part of Africa, while the lower ranks are located on the northern and southern extremes of the continent. In the case of the Asian continent, there were substantial spatial variations in the GII rankings. The average ranking was 74, with a standard deviation of 42. Thus, Asia had some of lowest rankings (indicative of lower gender inequality) in China (37) and Malaysia (38), along with the highest rankings (indicative of higher gender inequality) located in Yemen (152) and Afghanistan (149). Most of South Asia had GII ranks above 100, while the lower ranks were clustered in Central and East Asia, and the Middle East. Additionally, in general, countries that are landlocked are also usually the ones with greater gender inequality in the Global South.

There were substantial differences in the spatial patterns of GII and GDI rankings, particularly across the Africa and Asia. However certain countries were ranked higher for both indices, which include most of South Asia and Central Africa. Additionally, Bolivia in South America was also ranked higher for both indices. On the contrary, relatively lower ranks for both indices were clustered mostly in South America and the Caribbean. Thus more equal access to resources and rights among males and females can be assumed in this part of the Global South.

Global Gender Gap Index (GGGI)

Another measure that can be helpful for understanding the spatial patterns of gender inequities and inequalities is the Global Gender Gap Index (GGGI) published annually by the World Economic Forum as the Global Gender Gap Report (Schwab et al. 2014). It is an excellent source for monitoring progress or change in gender gap at the global level. One of the basic arguments for gender equality is that women constitute about half of the global population, and thus deserve equal access to resources, health, education, income, and representation. Despite repeated arguments about reducing gender inequality in different countries, there exist substantial disparities in various measures used to assess gender gap. The GGGI is developed using a four step process. First, all data are converted to male/female ratios. Some of the specific variables include literacy levels, enrollment rate, wage equality, female legislators, sex ratio at birth, and participation in parliament classified based on gender. The data are converted into ratios to capture the gaps between the attainment levels of males and females. Next, the ratios are truncated at the "equality benchmark", which is considered to be 1 for most of the variables, depicting equality between men and women. This index reveals the gaps in the outcome variables rather than the input variables and the ranks are based on overall gender equality, and not just women's empowerment. The gender based gaps are defined as the unequal access to the resources and opportunities in countries, and not the actual levels of resources and opportunities in different countries. Thus the GGGI is independent of the levels of development in each individual country.

It is evident from Fig. 2.4, not all countries in the Global South were included in the report. In order for a country to be included in the GGGI ranking, there had to be data available for at least 12 indicators of the 14 that constitute the index. The majority of the countries that were not included in the GGGI rankings are from Global South, located mainly in Central Asia and Africa. The average value for the overall rank for GGGI among the countries of the Global South was 83. The lowest rank in the Global South is located in Nicaragua (7) followed by Rwanda (7), while the highest rank was located in Yemen (142) followed by Pakistan (141). In the case of the overall scores, the average value was 0.68, with the lowest score located in Yemen (0.51) followed by Pakistan (0.55), while the maximum score is located in Nicaragua (0.7894) and Rwanda (0.7854). The scores can be interpreted as the gender gap in Pakistan and Yemen is at 50%, while it is almost at 80% in Rwanda and Nicaragua. Overall, there is almost a horizontal latitudinal gradient in the spatial patterns of the GGGI ranks and scores, with highest ranks associated with the lowest

Global Gender Gap Index (GGGI)

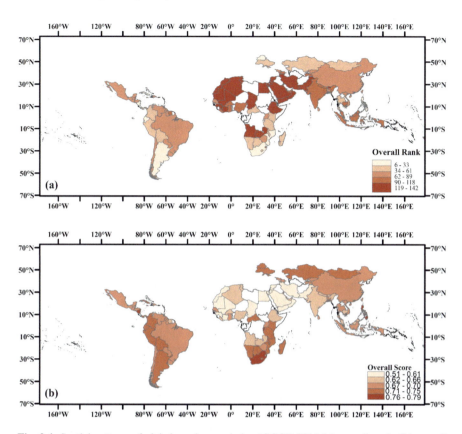

Fig. 2.4 Spatial patterns of global gender gap index (GGGI) 2014 (**a**) overall rank; (**b**) overall score

scores located closer to the equator. Additionally, the lowest scores were also more concentrated across countries in Asia and Africa.

However, there are substantial spatial variations in the overall rankings of the different countries at the regional scale. In Asia only three countries were ranked in the top 50 GGGI rankings, which include the Philippines (9), Mongolia (42), and Kazakhstan (43). It is noteworthy that two of the most populous nations, China and India, which have undergone rapid economic development in the recent years, were ranked 87 and 114 respectively. This gap is more evident in the case of education and empowerment indices, which are discussed later. It is particularly significant since both countries are also in the frontiers of adverse impacts of climate change, in the form of increasing average temperatures, and extreme events such as heavy rainfall and heat waves. Furthermore, the capital cities of China (Beijing) and India (New Delhi) have recorded the worst levels of particulate matter in the last 5 years. Additionally, all of the Middle Eastern nations were ranked above 100 in terms of the overall GGGI index, with the largest gender gaps observed in Yemen (142) followed by Lebanon (135), Jordan (134), and Saudi Arabia (130). The larger gaps in

Asia and the Middle East are mainly attributed to lower levels of empowerment and education for women. The Asia Pacific region performed quite well in terms of closing their overall gender gaps to 68%. Among the 18 countries which constitute this region, 11 of them showed an overall improvement in their scores. Most of the improvements in this region can be attributed to educational attainment and health and survival rates. On the contrary, scores decreased for the political empowerment sub-index. Furthermore, Kazakhstan in Central Asia fully closed the gender gap in two sub-indices of the GGGI, which include educational attainment and health and survival. However, three countries in Central Asia (Azerbaijan, Albania, and Armenia) had the lowest ranks in terms of health and survival. Armenia also had the lowest female to male sex ratio at birth in the world.

In general the largest gaps were concentrated across Northern Africa, with most of them observed in economic participation and opportunity sub-index. The index was not available for several countries across Africa including Egypt, Sudan, South Sudan, Gabon, Cameroon, and Equatorial Guinea. Additionally, there was a north south gradient. The largest gender gaps in Africa were observed in Chad (140) and Mali (138), which were also among the highest ranks in the entire Global South. These spatial patterns are similar to those observed in the case of the GII and GDI discussed above. However, the lowest GGGI ranks were observed in Rwanda (7), Burundi (17), and South Africa (18). About half of the countries in Africa were ranked in the top 50 globally. Therefore, in the African continent there were widespread variations in the overall GGGI rankings.

In the case of the Latin America and the Caribbean region the gender gap narrowed by 70%, exhibiting a steady improvement over the years. The narrowest gap was observed in the case of the health and survival index. This region also performed the best in term of narrowing the gender gaps since the index was formulated in 2006, with Guatemala showing the maximum improvement. As mentioned before, Nicaragua is ranked sixth at the global level and ranked third in the region in terms of overall improvement since 2006 (after Guatemala and Ecuador). The average ranking of GGGI for all of Latin America and the Caribbean was 59 with a standard deviation of 24.8, which is the lowest among all regions. Thus this region showed the minimum spatial variations in terms GGGI rankings. The highest GGGI rankings were located clustered over Suriname (109), Belize (100), and Guatemala (89).

Based on the spatial patterns revealed by the three indices on gender gap and inequality, some overall consistent patterns are revealed. All three indices indicate higher levels of gender inequality and gap in South Asia and North and Central parts of Africa. Countries that are landlocked in Asia and Africa also usually experienced higher levels of gender inequalities. Specifically some of the countries that consistently revealed higher gender gaps and inequality included Afghanistan, India, Pakistan in South Asia; and Chad, Niger, Mali in Africa. However, in the case of South and Central America and the Caribbean, only Bolivia consistently ranked high on gender inequality issues. Other countries in this region that showed relatively higher rates of gender gaps and inequalities include Guatemala and Honduras. In order to better understand the role of the various factors on the final ranks of the three indices discussed above, a more detailed discussion is needed. The various

Education 39

variables included in the calculation of the different indices to represent gender gap and inequality can be mainly classified under the three groups described below.

Education

Access to education and educational attainment is a robust indicator of gender equality, specifically for younger age children. Access to good education will lead to greater awareness among people, particularly among women from a young age. In a lot of societies gender based differences in access to opportunity start at a much earlier stage for children, which then become difficult to get rid of in adulthood. While equal access to educational opportunities is taken for granted in most parts of the more developed world, it is not the same in large sections of the Global South. Restrictions on girls' education ranges from limited to complete bans, as well as societal stereotypes that encourages girls to pursue certain lines of educational attainment. For instance, many girls from a young age are encouraged to go for less technical courses of study so that they can be engaged in less demanding vocations in their adult lives. While it is not wrong to choose any kind of vocation, encouraging gendered stereotypes can hinder progress toward gender equality. Education can be considered as the first step toward achieving gender equality from a young age. An educated woman has a greater chance of achieving economic and social equality, while decreasing the chance of dying during pregnancy and raising healthier children.

Some of the education variables included in the Human Development Report compiled by the UNDP annually include mean years of schooling, expected years of schooling, and population with at least some secondary education aged 25 and above. All of these indicators of education are available for males and females separately, and are used in the calculation of GDI and GII. The spatial patterns of each of the indicators are shown in Fig. 2.5.The data were not available for all of the countries in the Global South. The expected years of schooling for females were slightly higher than the mean years of schooling for majority of the countries (Fig. 2.5a). Overall, South America and the Caribbean show higher levels of education among females. For instance the average scores for Latin America and the Caribbean for mean years of schooling were 7.7, compared to 6.8 for East Asia and the Pacific, 3.7 for Sub-Saharan Africa, and 3.5 for South Asia. Specifically, the mean years of schooling for females in some of the countries in Sub Saharan Africa, Chad, Mozambique, Guinea and Niger, are less than 1 year.

In South America and the Caribbean region, there was little variation for the different sub-indices of education. The highest values for the mean years of schooling were observed across the Caribbean in the Bahamas (11.13), Trinidad and Tobago (10.91), and Cuba (10.05) (Fig. 2.5b). In South America, Argentina was ranked the highest at 10.02 years for mean years of schooling and 17.5 years for expected years of schooling. The expected years of schooling were also higher in the Caribbean region compared to the South American continent. The average for the mean years

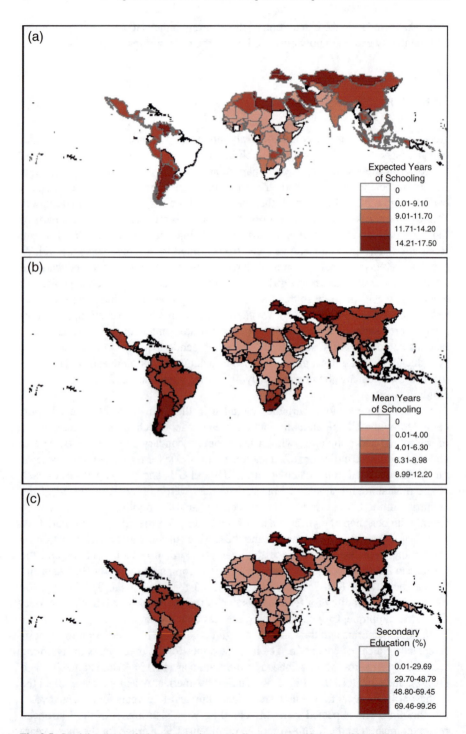

Fig. 2.5 Spatial patterns of education attainment indicators among females 2014 (a) Expected years of schooling; (b) Mean years of schooling; (c) Secondary Education for females aged 25 and above

Education   41

of schooling over the African continent was substantially lower at 3.7 years, with a maximum of 9.8 years in South Africa. This is substantially lower than the rest of the Global South. In the case of Asia, there was almost a bimodal distribution with the high values for mean years of schooling located across most of Central and East Asia, and the Pacific islands, while the lower values were concentrated in South Asia, including India (3.22) and parts of the Middle East, like Yemen (1.23). It is noteworthy that the countries with the highest mean years of schooling were also located in South Asia, Sri Lanka (13.9), and the Middle East, United Arab Emirates (10.15) and Qatar (10.09). The mean years of schooling for most of the Pacific Island nations were above 10 years.

Box 2.2 No Ceilings Report by Clinton Foundation
Recently the Clinton Foundation published a report entitled *"No Ceilings: the Full Participation Project"*. The main goal of this report was to inspire and advance the full participation of women and girls around the world. It is mainly a data driven evaluation of the progress made by women and girls toward equal access, representation, and compensation in communities around the world. The report highlighted the critical importance of gender equality on all communities, economies, and wider society. The report consists of a compilation of data assessing the gaps and progress made in eradicating gender inequality around the world since the UN Fourth World Conference in Beijing in 1995. The report showed that there are better efforts in data collection on women and girls around the world. However, there were still gaps in comprehensive data collection in certain areas, including women's economic contributions such as unpaid work at home, under reporting of violence against women, number of women living in poverty, and environmental risks faced by women. The report emphasizes the critical importance of human rights to women, which would guarantee autonomy in family and civic life, as well as access to quality education and health opportunities. One of the specific focus areas of the report is the disproportionate effects of environmental challenges, like climate change and natural disasters on women and girls. Specifically after major natural disasters, such as hurricanes and floods, women may become more susceptible to domestic violence as a result of increased alcohol consumption and the breakdown in overall law and order situation. However, it is hard to actually gauge the effects of climate change on women because most environmental data are not gender segregated. Therefore, it is more difficult to assess the actual impacts of climate change on women and girls.

While the mean and expected years in school is an important indicator of girls going to school at an early age, it is also important to assess the number of females who are able to remain in school to study up to secondary education levels and higher as shown in Fig. 2.5c. The spatial patterns of the percentage of females who have attended secondary education above the age of 25 are very similar to those observed of the mean years of schooling. The lower levels of females with secondary education were mostly concentrated in parts of North and Central Africa and South Asia. The overall mean percentage across continental Africa was only 18.6% for female population above 25 years of age with secondary education. Only four countries in Africa continent had more than 50% of the female population with secondary education, including Botswana (73.6%), South Africa (72.75%), Libya (55.6%), and Gabon (53.8%). Most of South America and the Caribbean had a proportion of females with secondary education greater than 50%, except Guatemala (21.9%), Haiti (22.5%), and Honduras (28%). In Asia, there is again a clear dichotomy between South Asia with substantially lower percentage of females with secondary education compared to East Asia and the Pacific. The lowest levels were again located in Afghanistan (5.8%), Papua New Guinea (6.8%), and Yemen (7.6%). The number of women with secondary education greater than 50% was found in most of Central and East Asia as well as the Middle East. Thus there is a clear positive correlation between education and gender inequality. Similarly educational attainment ranks calculated by the World Economic Forum for the Global Gender Gap Report also revealed similar spatial patterns with the higher levels of educational attainment in South America and the Caribbean, while the lowest levels were observed over parts of Africa and South Asia. Unfortunately these data were not available for most of the countries in continental Africa.

Empowerment

Empowerment is widely considered as a key indicator of gender equality. It indicates the amount of involvement women have in the decision-making processes through participation in the political process (by simply voting or through active involvement). Empowerment is not only limited to political processes but also gainful employment at all levels of various organizations. Overall, there is gradually greater visibility of women in managerial positions in major organizations, which is a direct result of better education opportunities availed by girls at a younger age. However, this is not the case in all countries of the world, particularly in certain countries of the Global South, where women do not have equal rights, thus creating major imbalances in the society. Not many nations in the world have had a popularly elected female leader, though there are quite a few in the Global South, including Ms. Chandrika Kumartunga in Sri Lanka, Ms. Indira Gandhi in India, Ms. Benazir Bhutto in Pakistan, Ms. Michelle Bachelet in Chile, Ms. Dilma Rousseff in Brazil, and Ms. Cristina Kirchner in Argentina. It is noteworthy that some of the countries with the highest level of gender inequalities in South Asia such as India (Indira Gandhi) and Pakistan (Benazir Bhutto) had popularly elected female leaders from prominent political families, both of whom were assassinated. Other countries in South and Southeast Asia with female leaders include the Philippines (Corazon

Aquino) and Bangladesh (Khaleda Zia and Sheikh Haseena Wazed), of whom all belong to powerful political families.

Box 2.3 Missing Women in Asia

Professor Amartya Sen published an editorial in BMJ in 1992 entitled "Missing Women: Social inequality outweighs women's survival advantage in Asia and North Africa", which highlighted the role of social factors on the significantly low female: male ratio. Female infanticide has also been referred to as "gendercide", used to describe the extermination of persons of a particular sex. Estimated that over 25 million women in India could potentially be alive today, but are not alive mainly due to sex selection at birth and mistreatment of young girls. Furthermore, estimates 163 million missing women due to female infanticide and child neglect resulting from poor nutrition in all of Asia. The countries with the highest proportion of missing women are clustered in South Asia including Pakistan, India, and Bangladesh. Social and cultural factors in these countries stem from patrilineal inheritance in patriarchal societies, coupled with a reliance on male children to provide economic support indefinitely.

In an update in 2003, Professor Sen raised concerns about sex- specific female fetus abortions prevalent in many countries of Asia, where there is a preference for the male child. He also highlighted the importance of spatial differences in the sex ratio at birth which is not represented in the overall rates reported for the country as a whole. For instance in north and west India, sex specific female fetus abortion is particularly widespread. One of the major factors cited for the higher age specific mortality rates among females in India was attributed to higher rates of maternal mortality rates. The prevalence of female infanticide, leading to lower birth ratios, as well as the more long term neglect of female health and nutrition particularly during childhood, were identified as some of the main factors. Professor Sen called for change in public policy to "rescue these missing women".

However, a recent article discussed the reduction in the number of unwanted and dying girls due to their reduced wellbeing and pre-natal sex detection during 1980s and early 1990s in Taiwan. They called for the formulation of policies incentivizing parents to invest in daughters. Similar results were also found in India: the reduction in malnutrition in the form of underweight and wasting in response to increase in prenatal sex selection.

Various indicators can be used to measure the level of empowerment among women as an indicator of gender equality. Based on data availability for various countries in the Global South, two indicators have been used to assess the spatial

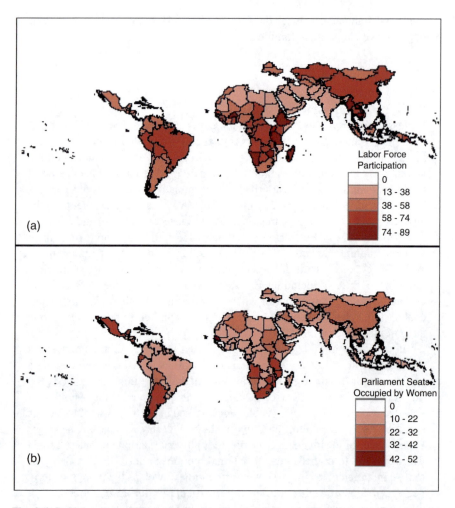

Fig. 2.6 Spatial patterns of empowerment indicators (**a**) Female labor force participation; (**b**) Parliament seats occupied by women

patterns of women's empowerment, female labor force participation and share of seats in parliament (Fig. 2.6). The spatial patterns for both indicators are quite similar across Central, East, and Southeast Asia, where the relatively higher levels of female participation in the work force or political processes are mostly concentrated. On the other hand, consistently lower patterns were observed in the countries of South Asia, the Middle East, and Northern and Sub Saharan Africa. In general, the rates of female participation in political processes and labor force were relatively higher with increasing distance from the center of the Global South, in the Middle East and South Asia, toward the south or north, including Chile and Argentina in South America, and South Africa, Mozambique, and Namibia in Africa. In South America and the Caribbean, women occupy 25% of the parliamentary seats, while it is less than 14% in the Arab States. It is important to note that empowerment is directly related to educational attainment, which is why in some countries female education is constrained as evidenced by the broad daylight shooting of Malala Yousafzai in Pakistan. Additionally, empowerment is one of the more long term paths to removing gender inequality. Increasing women's empowerment through increased participation in the economy and political processes leads to a decline in poverty levels among women.

Health and Survival

The social norms in many of the countries in the Global South are based on a preference for a male child. This has resulted in a long term imbalance in the sex ratio, with fewer female children born. The majority of the countries in South Asia and Sub Saharan Africa have a very low sex ratio, meaning fewer female children born compared to males. The Global Gender Gap Report labels this as the "missing women" issue, proposed by noble laureate Professor Amartya Sen in 1992 (Sen 1992). A recent study, highlighted the negative impacts due to the preference for not only the male child, but more specifically the eldest male child in poorer families with limited resources in India (Jayachandran and Pande 2015). They analyzed data for 174,000 Indian and Sub-Saharan African children and found that the Indian first born male children were significantly taller than their siblings after them. Comparatively, the heights of girls who were the eldest in their families were found to be the most stunted in terms of height. Other than the birth rate and sex ratio, there are also disparities in the access to proper health care for women in the Global South. However, there is a general agreement that health and nutrition for women and girls have improved over the years, though much needs to be done in order to achieve long term resilience. As highlighted above, education is a key factor for better heath among women, it has been reported that most of the reduction in child mortality between 1970 and 2009 can be attributed to increased educational attainment (Karlsen et al. 2011; Semba et al. 2008; Strauss and Thomas 1995; Gakidou et al. 2010; Bbaale 2015).

Therefore, two indicators have been used to assess the quality of health and survival in terms of gender disparities, which include maternal mortality rates (MMR) and female life expectancy (Fig. 2.7). Overall, MMR have decreased globally due to improved access to health services (Fig. 2.7a). Better access to health care services, along with higher educational attainment has led to a drop in MMR and child mortality rates. However, certain countries in the Global South continue to experience relatively higher MMR, such as the rates per 100,000 live births observed in Chad (1100), Somalia (1000), and Sierra Leone (890). Specifically, 19 of the countries with the highest MMR were located in Sub Saharan Africa. Additionally, in Mali and Swaziland the male life expectancy was higher than female life expectancy. In the case of Asia, the higher MMR were again clustered in South Asia, with the highest rates observed in Laos (470), Afghanistan (460), and Pakistan (260).

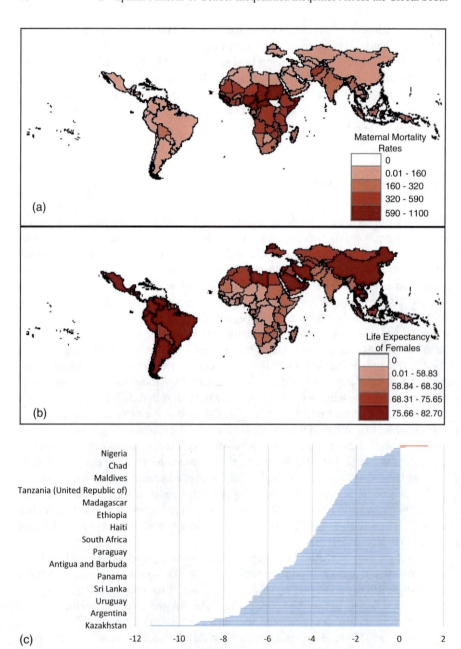

Fig. 2.7 Spatial patterns of health and survival indicators (**a**) Maternal Mortality Rate per 100,000 live births; (**b**) Life Expectancy Rate; (**c**) Difference between male and female life expectancy (male-female)

The MMR were much lower across most of East and Central Asia, which can be attributed to better health infrastructure. They were mostly lower than 100 in South and Central America and the Caribbean; only Haiti (350), Guyana (280), Bolivia (190), Dominican Republic (150), Suriname (130), and Ecuador (110) experienced more than 100 deaths. Although, MMR is only a female specific indicator, it is an important indicator of development, mainly in view of the potential serious implications of maternal mortality on new born babies and their older siblings. Additionally, adolescent births could also be debilitating, given that 110 births per 1000 births were by women aged between 15 and 19 years in Sub Saharan Africa (UNDP 2015a, b, c, d).

Female life expectancy accounts for years lost due to deadly diseases as well as violence on women (Fig. 2.7b). In 2012 it was found that women lived longer than men at 73 years versus 69 years in 2012 (World Bank 2015). Additionally, there has been a greater improvement in the longevity of females in developing countries of the Global South. For instance, life expectancy for women in Ethiopia rose from 51 to 65 years from 1995 to 2012 (Fig. 2.7b). There has also been an overall decline in the rates of deaths due to infectious and other deadly diseases all around the world (Center for Disease Control 2015). Furthermore, the global mortality rate for girls under 5 has also declined, with some of the highest declines observed over South Asia (57%) and Sub Saharan Africa (49%). The spatial patterns of the life expectancy across the Global South reveals lower life expectancy of less than 60 years in Central and South Africa, followed by 58–78 years in Eastern Africa and South Asia. The countries with the lowest female life expectancies include Sierra Leone (45.8), Swaziland (48.29), Lesotho (49.5), and Mozambique (51). The lower life expectancy rates in large parts of Africa are mainly due to early deaths caused by the prevalence of HIV/AIDS, which has shown improvement as a result of antiretroviral therapy (World Health Organization 2015). Other leading causes of deaths include malnutrition, diarrhea, and malaria. The average female life expectancy across most of the Caribbean and South America is above 75 years, except the two landlocked countries of Bolivia (69.5 years) and Paraguay (74.6 years), which showed lower female life expectancy. Moreover, when the female life expectancy was compared with the male life expectancy, only two countries, Mali and Swaziland showed lower life expectancy compared to the male life expectancy (Fig. 2.7c). Most countries showed very little difference between male and female life expectancy, which is contrary to the normal expectation of higher female life expectancy.

Climate Vulnerability Index

Based on the discussions above it is clearly evident that there exist significant gender inequalities and gaps at the regional scale. A selected set of representative factors have been analyzed and mapped in the previous sections to bring out the gender differences at the country level. While there are differences in the spatial patterns for the different variables taken into consideration, certain countries/regions consistently show up as the areas with higher gender inequalities, which include South Asia, and Sub Saharan Africa, and parts of the Middle East particularly Yemen. In a recently published article, the gendered implications of climate change, specifically in South Asia were emphasized, where patriarchal norms, inequities, and inequalities often make women more vulnerable (Sultana 2014). Similar dynamics are also present in the Middle East, leading to wide gender disparities. The processes are more complex in Sub Saharan Africa, where other than societal norms the impacts of climate change may also indirectly impact resource allocation, resulting in conflicts in several regions. This is particularly important in view of the already occurring and projected impacts of climate change in the Global South. Therefore, it is pertinent to analyze the spatial relationship between gender inequalities/gaps and climate change vulnerabilities.

There are methods to gauge climate change vulnerability. Among them, The Notre Dame-Global Adaptation Index (ND-GAIN) is an open-source index that reveals the vulnerability and exposure to climate change impacts at the country level (University of Notre Dame 2015). It also assesses the readiness for different adaptation actions that can be taken by both public and private sector investment in different countries. Vulnerability levels of each country are measured by its exposure, sensitivity, and capacity to adapt to the impacts of climate change. It specifically takes into consideration six "life-supporting" (University of Notre Dame 2015) factors, which include food, water, health, ecosystem services, human habitat, and infrastructure. All of these factors are based on the geographic location and socio economic conditions in each country. Some of these specific factors, such as water availability and food security will be explored in greater detail in later chapters. Thus, the exposure and sensitivity of each country is computed by the degree to which they are exposed to or responsive to climate change related processes. The adaptive capacity of each country is estimated by the availability of resources to put adaptation in place in order to reduce vulnerability and exposure to climate change. The readiness of each country is determined by levels of economic, government, and social readiness. However, not all countries are well prepared to address the impacts of climate change at the local level, which is assessed by the readiness index part of ND-GAIN index. It measures each individual country's capability to apply economic investments and convert them to adaptation actions (University of Notre Dame 2015). It consists of three measures of overall readiness: economic readiness, governance readiness and social readiness.

The spatial patterns of the vulnerability and readiness indices have been mapped in Fig. 2.8. In general most of South America and Asia (except South Asia) have experienced a lower vulnerability index along with higher levels of readiness in terms of their preparedness for the impacts of climate change. Most Sub Saharan African countries experienced the highest levels of vulnerability to climate change processes, such as increased intensity of droughts, water scarcity, and health impacts (e.g. spread of infectious diseases). However, certain countries, such as Rwanda in Sub Saharan Africa, have been able reduce their vulnerability to climate change

Climate Vulnerability Index

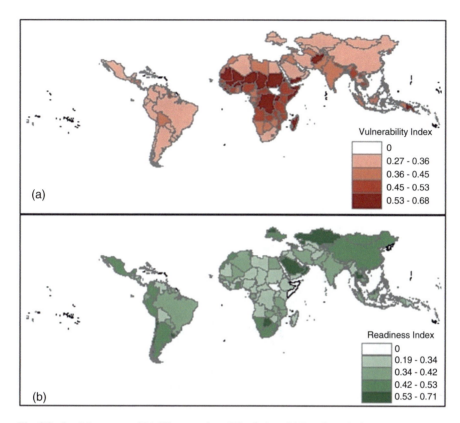

Fig. 2.8 Spatial patterns of (**a**) Climate vulnerability index; (**b**) Readiness index

impacts through the improvement in health indicators such as reduced malnutrition among children and slum populations. Impacts from sea level rise have also made some of the Pacific islands (such as the.

Solomon Islands, and Papua New Guinea) rank very high in terms of the vulnerability index. In Asia and the Middle East, Afghanistan (0.58), Bangladesh (0.5), and Yemen (0.55) showed the highest vulnerability. Bangladesh is a low lying country in South Asia, and is already facing the impacts of climate change as a result of sea level rise and intense extreme weather events such as hurricanes. It is also noteworthy that the areas which are most vulnerable to climate change, which include South Asia, Sub Saharan Africa, and parts of the Middle East, such as Yemen, are also areas with relatively high levels of gender inequalities/gaps. Additionally, Bolivia, which consistently ranked high in gender inequality indices also ranked high in climate vulnerability index, but not in the exposure index. Both the vulnerability index and readiness index range from 0 to 1. Higher values of the vulnerability index represent higher levels of vulnerability and exposure to climate change impacts. While higher values for the readiness index represents greater preparedness to already occurring and impending impacts of climate change. Thus, a preliminary analysis reveals regions with relatively higher gender inequalities are also more vulnerable to impacts of climate change.

> **Box 2.4 Role of Mary Robinson Foundation Climate Justice in promoting the development of gender-informed climate policy**
>
>
>
> The Foundation engaged in contributing to the development of gender-informed climate policy. They have set out commitments to address the gender dimensions of climate change and provide guidance on how best to do this. They also aim to strengthen references to gender and gender equality and women's leadership in international policy, thus facilitate more gender responsive action on the ground. Some of the key areas the foundation is working on are listed below:
>
> 1. Strengthen references to gender equality and women's leadership in international climate policy in order to facilitate more action on the ground.
> 2. Contribute to the consolidation of women's leadership in the fields of climate change, gender and sustainable development to highlight the gender dimensions of climate change
> 3. Increase the impact of women's leadership for policy influence by providing a sound evidence base and links to the grassroots.
>
> The foundation has established Women's Leadership on Climate Justice Network, as a communication network of grassroots and grassroots-linked organizations working on issues of gender and climate change.

There is growing awareness about the greater adverse impacts of climate change on women and girls, as evidenced in the special sessions during recent Conference of Parties (COP) meetings, as well as abundant regional case studies literature published by various non-governmental organizations. Most of the literature is based on local case studies, which have been produced by Gender and Development (GED), the Consultative Group for International Agricultural Research (CGIAR), and various agencies under the UN (UN Fund for Women, UNDP, UN Women, and UN Environment Program). Thus as a result of cultural and social norms, women and girls, particularly in the Global South, will be more exposed to the negative impacts of climate change. Some of the key recommendations made in the literature are listed below:

- There is a critical need for gender sensitive climate policy formulation from the grass roots level.
- Women should not be only regarded as victims of climate change, but rather as drivers and sources of knowledge for effective mitigation and adaptation policies.

- Women's participation in the policy making process should be increased through greater representation at all levels of policy formulation.
- Gender related issues should be incorporated in discussions with stakeholders at different levels.

However, there is limited academic literature highlighting the gendered differences in the impacts of climate change, particularly in the Global South. Over the last few years several scholars have called for the need to incorporate gender in not only climate change literature but also policy formulation. For instance, a more in-depth gender analysis has been argued for, which should include critical feminist theorizing of processes that shape today's climate politics (MacGregor 2010). It is widely validated in both academic literature and mainstream media that climate change is one of the greatest environmental threats of all time. The impacts are not gender neutral, but rather women are disproportionately disadvantaged by the impacts of climate change. Therefore, gender mainstreaming is crucial for an efficient adaptation to the impacts of climate change. This can be achieved through more in-depth analysis of both direct and indirect impacts of climate change.

References

Agostine, A., & Lizarde, R. (2012). Gender and climate justice. *Development, 55*, 90–95.

Bardhan, K., & Klasen, S. (1999). UNDP's gender-related indices: A critical review. *World Development, 27*, 985–1010.

Bbaale, E. (2015). *Female education, labour-force participation and fertility: Evidence from Uganda*. Retrieved August 5, 2015, from http://www.csae.ox.ac.uk/conferences/2011-EDiA/papers/847-Bbaale.pdf

Brody, A., Demetriades, J., Esplen, E. (2008). *Gender and climate change: Mapping the linkages: A scoping study on knowledge and caps*. Brighton, UK: BRIDGE.

Center for Disease Control. (2015). *Achievements in public health, 1900-1999: Control of infectious diseases*. Retrieved July 8, 2015, from http://www.cdc.gov/mmwr/preview/mmwrhtml/mm4829a1.htm

Clinton Foundation. (2015). *No ceilings: The full participation project*. Retrieved July 1, 2015, from http://noceilings.org/

Dankelman, I. (2010). *Gender and climate change: An introduction*. London: Earthscan.

Denton, F. (2002). Climate change vulnerability, impacts, and adaptation: Why does gender matter? *Gender and Development, 10*, 10–20.

FAO. (2015). *Women and sustainable food security*. Retrieved June 19, 2015, from http://www.fao.org/docrep/x0171e/x0171e02.htm

Gakidou, E., Cowling, K., Lozano, R., & Murray, C. J. L. (2010). Increased educational attainment and its effect on child mortality in 175 countries between 1970 and 2009: A systematic analysis. *The Lancet, 376*, 959–974. https://doi.org/10.1016/S0140-6736(10)61257-3.

Jayachandran, S., & Pande, R. (2015). Why are Indian children so short?. *HKS Faculty Research Working Paper Series RWP15-016*.

Karlsen, S., Say, L., Souza, J. P., Hogue, C. J., Calles, D. L., Gülmezoglu, A. M., et al. (2011). The relationship between maternal education and mortality among women giving birth in health care institutions: Analysis of the cross sectional World Health Organization global survey on maternal and perinatal health. *BMC Public Health, 11*, 606–606. https://doi.org/10.1186/1471-2458-11-606.

MacGregor, S. (2010). 'Gender and climate change': From impacts to discourses. *Journal of the Indian Ocean Region, 6*, 223–238. https://doi.org/10.1080/19480881.2010.536669.

Schwab, K., Eide, E. B., Zahidi, S., Bekhouche, Y., Ugarte, P.P., Camus, J., et al. (2014). The global gender gap index 2014. In *World economic forum*. Berkeley: Harvard University and the University of California. http://reports.weforum.org/global-gender-gap-report-2014/part-1

Semba, R. D., de Pee, S., Sun, K., Sari, M., Akhter, N., & Bloem, M. W. (2008). Effect of parental formal education on risk of child stunting in Indonesia and Bangladesh: A cross-sectional study. *The Lancet, 371*, 322–328. https://doi.org/10.1016/S0140-6736(08)60169-5.

Sen, A. K. (1992). Missing women. *BMJ, 304*, 586–587.

Strauss, J., & Thomas, D. (1995). Chap. 34: Human resources: Empirical Modeling of household and family decisions. In *Handbook of development economics* (Vol. 3, pp. 1883–2023). New York: Elsevier.

Sultana, F. (2014). Gendering climate change: Geographical insights. *The Professional Geographer, 66*, 372–381. https://doi.org/10.1080/00330124.2013.821730.

Terry, G. (2009). No climate justice without gender justice: An overview of the issues. *Gender and Development, 17*, 5–18.

UN. (2015). *SDGs: Sustainable development knowledge platform, United Nations*. Retrieved December 29, 2016, from https://sustainabledevelopment.un.org/sdgs

UN Population Fund. (2015). *State of world population 2009 facing a changing world: Women, population and climate*. Retrieved July 1, 2015, from http://www.unfpa.org/sites/default/files/pub-pdf/state_of_world_population_2009.pdf

UN Women. (2015). *Facts & figures*. Retrieved July 1, 2015, from http://www.unwomen.org/en/news/in-focus/commission-on-the-status-of-women-2012/facts-and-figures

UNCCD. (2015). *The forgotten billion MDG: Achievement in the drylands*. Retrieved August 5, 2015, from http://www.unccd.int/Lists/SiteDocumentLibrary/Publications/Forgotten%20Billion.pdf

UNDP. (2015a). *Gender development index*. Retrieved April 16, 2015, from http://hdr.undp.org/en/content/gender-development-index-gdi

UNDP. (2015b). *Gender inequality index*. Retrieved April 16, 2015, from http://hdr.undp.org/en/content/gender-inequality-index-gii

UNDP. (2015c). *Human development report 2014 - Sustaining human progress: Reducing vulnerabilities and building resilience*. Retrieved August 5, 2015, from http://hdr.undp.org/sites/default/files/hdr14-report-en-1.pdf

UNDP. (2015d). *Resource guide on gender and climate change*. Retrieved June 19, 2015, from http://www.undp.org/content/dam/aplaws/publication/en/publications/womens-empowerment/resource-guide-on-gender-and-climate-change/Resource.pdf

UNESCO. (2015). *Facts and figures: Violence against women and girls*. Retrieved July 1, 2015, from http://www.unesco.org/new/en/communication-and-information/crosscutting-priorities/gender-and-media/women-make-the-news/facts-and-figures/

UNFPA. (2015). *Maternal health*. Retrieved July 1, 2015, from http://www.unfpa.org/maternal-health

University of Notre Dame. (2015). *Notre dame global adaptation index*. Retrieved June 16, 2015, http://www.gain.org/

WHO. (2015). *Gender, climate change and health*. Retrieved June 19, 2015, from http://www.who.int/globalchange/GenderClimateChangeHealthfinal.pdf

World Bank. (2015). *Life expectancy at birth, female (years)*. Retrieved August 5, 2015, from http://data.worldbank.org/indicator/SP.DYN.LE00.FE.IN

World Health Organization. (2015). *Life expectancy situation*. Retrieved June 11, 2015, from http://www.who.int/gho/mortality_burden_disease/life_tables/situation_trends_text/en/

World Health Organization and United Nations Children's Fund. (2015). *Progress on drinking water and sanitation: 2012 update*. Retrieved July 1, 2015, from http://www.unwomen.org/en/what-we-do/economic-empowerment/facts-and-figures#sthash.tAe13kAG.dpuf

Chapter 3
Health

"Communities and countries and ultimately the world are only as strong as the health of their women."

—Michelle Obama

Sustainable Development Goal 3: "Ensure healthy lives and promote well-being for all at all ages" (UN 2016).

Introduction

Some of the specific targets within the third overall goal under the recently adopted United Nations Sustainable Development Goals (SDGs) include the end of epidemics, like tuberculosis, AIDS, malaria and neglected tropical diseases. These also include the reduction of premature mortality from non-communicable diseases by one third through prevention and treatment. In order to achieve the overall goal and attain the specific targets, the impacts of climate change need to be taken into consideration. There is increasing understanding and awareness about the direct and indirect impacts of climate change on communicable and non-communicable diseases. Also, the impacts of climate change on human health are not gender neutral. Specifically, in the Global South, the impacts are disproportionately greater on women's health than men, due to combination of existing health infrastructure, lack of women's empowerment, and cultural norms. The indicators for women's health should not be limited to only reproductive or maternal health, but should include the entire lifespan to address the needs of girls and older women (Horton and Ceschia 2015). This is important particularly for girls at a younger age, who may experience high levels of malnutrition as a result food shortage. Additionally, women constitute a major portion of the formal and informal workforce in the health sector. Their contributions are not completely accounted for in the economy. They are also the main caregivers at the family level, and thus exposed to diseases. Therefore a gender sensitive understanding of the impacts of climate change on human health is essential.

In this context, variability in climatic conditions has extensive impacts on human and physical systems at different spatial scales. These impacts have wide spatial variations, which are a function of the local environmental factors, topography, and in particular the vulnerability of the local population. There is widespread evidence of the negative impacts across the globe in the form of increased outbreaks of vector-borne diseases spreading both horizontally and vertically across the earth surface. For example, the aggravating impacts of changing and variable climate conditions on the frequency and intensity of commonly occurring infectious diseases such as flu, malaria, and dengue are well documented (McMichael et al. 2006). There is also an evidence of the spread of some of these infectious diseases, which are usually concentrated in the low lying areas of the tropics, to higher elevations in the tropics and higher latitudes as a result of warmer temperatures (Zhou et al. 2004). There is also an increase in the amount of time during which incidences of these diseases are being observed, particularly in the tropics. In view of rising average temperatures and predictions for higher frequency of heat waves, more people are and will be exposed to heat related stresses. This will result in greater number of heat related mortality in urban areas (McGeehin and Mirabelli 2001). The role of extreme weather events in the form of droughts, floods caused by extreme heavy precipitation or hurricanes, is becoming more evident on disease incidences particularly in the developing world. In previous chapters, the distinct spatial variations in the long term trends in climate change related processes have been highlighted. Therefore, the impacts of climate change on human health are also not uniform across the globe. There is clear evidence of the most severe adverse impacts of climate change on human health on the poorest and vulnerable populations of Global South (WHO and WMO 2015). Some of the key aspects of climate change and human health highlighted in scientific literature and leading organizations are described below:

- Processes related to climate change have widespread effects on the social and environmental determinants of health such as water supply, air quality, infrastructure, housing, food supply, and access to proper medical attention.
- The World Health Organization (WHO) estimates the direct damage costs to heath between US$ 2–4 billion per year by 2030.
- In 2000, climate change related processes caused 2.4% of worldwide diarrhea, and 6% of malaria in some middle-income countries (WHO 2015a).
- Extreme high air temperatures have directly resulted in deaths from cardiovascular and respiratory disease.
- Globally, the number of reported weather-related natural disasters has more than tripled since the 1960s. Each year, these disasters result in over 60,000 deaths, mainly in developing countries (WHO 2017).
- Variable rainfall patterns have a direct impact on the supply and contamination of freshwater leading to the spread of water borne infectious diseases.

Introduction

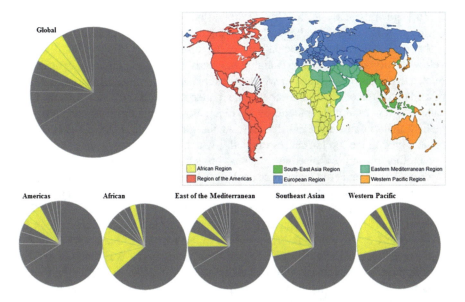

Fig. 3.1 Top ten leading causes of deaths at the global and regional scale in 2012. Yellow sections in the pie chart represent deaths that can be related directly or indirectly to climate change (WHO 2015c). The World Map on the top right shows the demarcation of the WHO regions. It is important to note that the entire Americas are grouped in one single region, which may not reveal the differences between the more developed and less populated north and the opposite in the South

All of these negative impacts are further aggravated for population inhabiting the low lying islands, coastal areas, mega cities, and mountainous and polar regions, most of which are located in the Global South. Therefore, it is important to focus on the spatial patterns of long term impacts of climate change on human health. In order to better understand these spatial patterns, we need to examine them at various spatial and temporal scales based on examining local exposures to more widespread exposures. For instance, local level impacts can be in the form of a natural disaster occurring in an area, resulting in the rapid spread of water borne diseases, such as cholera. On the other hand, the long term sustained increase in temperatures and GHGs will lead to more widespread and slower rate of impacts on human health.

Some of the direct impacts of climate change include the exposure to extreme heat or cold, increased levels of pollutants in the atmosphere, and extreme weather events. Indirect impacts include exposure in the form of spread of infectious diseases due to favorable conditions for the spread of vectors. However, there are substantial spatial variations in the magnitude of these impacts, the effects of which are further determined by the adaptive capacity of the local population. Therefore, the main focus of this chapter is to examine the impact of climate change on human health at various temporal and spatial scales, through a gender disaggregated analysis.

To begin this analysis, let's first examine how many of the top ten causes of deaths in 2012 were related directly or indirectly to climate. Figure 3.1 shows the global and WHO regional scale patterns of the top ten causes of deaths in 2012. The yellow sections in the pie-charts highlight deaths that may be directly or indirectly related to climate change and or weather patterns. Causes of deaths that are directly related to the effect of climate and weather patterns include trachea, bronchus, lung cancer, lower respiratory infections, and tuberculosis, while those related indirectly include malaria and diarrheal diseases. At the global scale three of these causes made it among the top ten causes of deaths, which include lower respiratory infections, other respiratory diseases (trachea, bronchus, lung cancers), and diarrheal diseases. However, at the regional scale the patterns are distinctly different. Overall, the highest proportion of climate related causes of deaths were concentrated in the African region, while the lowest were concentrated in the American region. The lowest proportion of deaths caused by climate in the American region may not be accurate for Central and South America, due to the merging of both North America (more developed) and South America (less developed) by the WHO. In the African region, three of the causes were ranked among the top five of the ten causes of deaths in 2012, which include lower respiratory infection (11.2%), diarrheal diseases (6.5%), and malaria (6.0%). As is shown in Fig. 3.1, the WHO African region excludes northern Africa in the African continent, but includes Sub Saharan Africa which experiences one of the highest levels of gender gaps. In the case of Southeast Asia, which ranked consistently high in gender inequality indices, three of the top ten causes of deaths in 2012 were related to climate change. These causes included lower respiratory infections (5.5%), diarrheal diseases (4.7%), and tuberculosis (3.3%). Some of the large metropolitan areas with the highest levels of air and water pollution, which include New Delhi, Bangkok, Dhaka, and Karachi, are located in this region. In the Western Pacific region, the trachea, bronchus, lung cancers (5.7%) and lower respiratory diseases (3.6%) were among the top ten causes of deaths. The dominance of respiratory related diseases in the top ten causes of deaths in this region is again related to poor air quality levels in some of the large metropolitan areas of Eastern Asia, including Beijing, Shanghai, and Manila. Finally, in the Eastern Mediterranean region, lower respiratory diseases (7.7%) and diarrheal diseases (3.5%) were ranked among the top five causes of deaths in 2012. Thus it is evident in most of the Global South climate conditions play an important role, particularly in Sub Saharan Africa where climate contributed to 23.7% of the total mortality. It is also important that most of the climate related deaths are related to air pollution and to some extent water pollution, which have only worsened over the years. Additionally, in view of the projected increase in GHGs, the situation will further deteriorate given the rapid rate of urbanization accompanied by declining air quality. However, in addition there will be significant localized non-fatal impacts of climate change on human health, which includes the spread of infectious diseases in the tropics. Therefore the main focus of this chapter will be on the impacts of climate variability and change on human health in the form of communicable and non-communicable diseases. Impacts of air and water pollution on human health will be discussed in greater detail in later chapters.

Infectious Diseases

According to the WHO, about 1/6th of the illness and disability worldwide can be attributed to vector-borne diseases, which puts substantial section of the world's population at risk. Each year more than one billion people are infected, out of which approximately one million people die from infectious vector borne diseases, including malaria and dengue (WHO 2008; Lozano et al. 2012). Infectious diseases are classified in two categories on the basis of their mode of transmission, person to person transmission through direct exposure, and indirect transmission through vector and non-biological mode, such as water and soil. Vector borne diseases have widespread socio-economic impacts, particularly in the Global South by increasing health inequities due to poor health infrastructure, particularly in rural areas. For instance, the per capita mortality rates of vector borne diseases are 300 times greater in developing nations (WHO 2008). This is mainly due to the prevalence of vector borne diseases in the warmer tropical climate of the Global South accompanied by lower levels of socio-economic development and inadequate health care support. Moreover the poorer sections of the society are more vulnerable to these diseases due to poor sanitary conditions, lower levels of awareness, and limited access to health care. Children under 5 are the most vulnerable population age group. For instance, an estimated 500,000 people are hospitalized every year with severe dengue, a significant proportion of who are children (Campbell-Lendrum et al. 2015). Vector borne diseases not only impact rural areas but also have severe impacts in large urban areas, such as recent dengue outbreaks in some of the large metropolitan areas of the Global South, including New Delhi, India.

The incubation time of a vector-borne infective agent within a vector organism is highly sensitive to the slightest changes in climate conditions in the form of fluctuations such as temperature, humidity, sunlight, and altitude (McMichael et al. 2003). Recently, in view of variable climatic conditions there have been increasing concerns about the impact of climate change on the spatial spread and temporal frequency of infectious diseases such as malaria and dengue. Modifying influences are not only limited to the ambient environment conditions, but also include human and physical conditions. For instance the effect of land use changes, in the form of urbanization, agriculture, deforestation play a significant role. The role of inter-annual and inter-decadal climate variability on the epidemiology of vector borne diseases is widely established (Githeko et al. 2000). Researchers predict the most significant effect on the transmission of vector borne diseases will be observed in the extremes of temperature ranges, which are 14–18 °C at the lower range and 35–40 °C at the upper range (Githeko et al. 2000). Increase in temperatures in the lower end has a greater impact on the transmission and incubation period (Watts et al. 1987). It is important to mention that extreme high or low temperatures are not favorable for the transmission and survival of disease-causing pathogens. However, a sustained increase in temperatures will lead to the increased tolerance among the vectors or the change in the spatial occurrence. In some cases these vectors will be able to evolve under the changing conditions or hosts switching to different parasites in changing environments (McMichael et al. 2003; Hoberg 2015). Specifically, mosquito species such as *Anopheles gambiae complex, A. funestus, A. darlingi, Culex quinquefasciatus,* and *Aedes aegypti* that are responsible for transmission of most vector borne diseases are sensitive to temperature fluctuations in immature and adult stages. Rise in water temperatures leads to faster maturation of mosquito larvae and thus greater production of offspring during the peak transmission period (Rueda et al. 1990). Additionally, in warmer climates female mosquitos digest blood faster and thus can feed more frequently (Gillies 1953). Therefore the vector transmission is accompanied by the shorter time extrinsic incubation period for malaria parasites and viruses within female mosquitos (Turell 1989).

Other than changes in temperature patterns, variation in the timing and spatial spread of precipitation also has an impact on the incidence of infectious diseases. Increased precipitation can result in the increase in the spread and quality of breeding sites for vectors like mosquitos, ticks, and snails. It can also lead to the development of more lush and dense vegetation, which are ideal resting sites for mosquitos. Similar to the negative impacts of extreme temperatures on the survival rates of vectors and parasites, extreme precipitation can also result in flooding and the run off of vector population. Overall humid conditions lead to the formation of standing pools of water which are favorable for mosquito breeding. On the contrary very dry conditions resulting from droughts can also lead to storage of water in containers, which are also breeding grounds for mosquitos (Patz et al. 2000). Additionally, there is increasing evidence of changing spatial patterns of diseases as a result of climate change, globalization, increased mobility, rapid urbanization, and land use land cover changes. In view of the importance of the relationship between vector borne diseases and climate variability, the following sections consist of a more detailed discussion of two of the most prevalent infectious vector borne diseases, malaria and dengue.

Malaria

Malaria is a parasitic disease, which is spread by the bites of female Anopheles mosquitos infected by malaria parasites such as *Plasmodium vivax* and *Plasmodium falciparum.* It is a major threat to human health in the developing world, mostly concentrated in the tropics. However, the spatial spread of malaria within the tropics is limited as observed due to their absence in high altitude and arid areas and in some cases successful control programs. Malaria is also limited temporally by its relative absence in cooler months. In general the highest incidence is found in Sub Saharan Africa and in parts of Oceania including Papua New Guinea (CDC 2015). Overall, the spatial patterns of malaria incidence are significantly limited by climatic conditions, including rainfall, humidity, and temperature, and the local capacity to deal with the disease. The range of temperatures determines how long mosquitos live, duration to adulthood, frequency of bites, and number of mosquitos. A substantial body of research on the impact of climatic variables on the spread of

Infectious Diseases 59

malaria and parasite development has accumulated since the mid-twentieth century (Chaves and Koenraadt 2010; Koenraadt et al. 2014; Béguin et al. 2011). Indeed, the relationship between the ecological components of the malarial cycle and temperature is seen to be the primary determinant in the occurrence of the disease. The influence of temperature on malaria development was found to be both non-linear and vector specific (Alonso et al. 2011). Thus when this factor shifts, in this case upwards with climate change, there will be consequences for malaria and those it affects (Bacaër and Guernaoui 2006). Fluctuations in temperatures can substantially alter the incubation period of parasites and its resulting transmission. Specifically, diurnal temperature fluctuation around means greater than 21 °C slows parasite development compared to constant temperatures (Paaijmans et al. 2009). In view of warmer climate predictions, the majority of the literature examining the impact of climate change on malaria transmission reveals its expansion into regions which are relatively cool at present (Matsuoka and Kai 1994; Martin and Lefebvre 1995; Martens et al. 1995; Martens et al. 1995). However, in a recently published study using a statistical approach to examine the spatial extent of malaria under projected climatic conditions showed limited changes in the spatial extent even under very extreme conditions (Rogers and Randolph 2000). However, the prevalence of malaria is mainly concentrated in the tropics which overlap with the core of the Global South. A later study consisting of multi-malarial model, multi model simulation data, and multi scenario comparison indicated significant uncertainty in the spatial extension of malaria. The most consistent increases were observed in the highlands of Africa and parts of South America and Asia (Caminade et al. 2014). Specifically in Africa, a shorter malaria season is predicted in West Africa with a longer transmission season predicted in East Africa under warmer conditions (Emert et al. 2013).

Box 3.1 Spread of Malaria in Higher Altitudes in the Global South

One of the consistent findings in recent research studies investigating the impacts of climate change on the spatial spread of malaria, is the increased risk of its spread in the higher altitudes. According to a recently published report in Science magazine by Siraj and coauthors in 2014, two regions that are particularly vulnerable to the risk of emergence of malaria include the highlands in Ethiopia in eastern Africa and Colombia. This is critical because the highland areas have so long been protected from this deadly disease, and therefore densely populated. Specifically, in Ethiopia almost half of the population live at altitudes between 1600 m and 2400 m. It is estimated that a 1 °C increase in temperature will lead to an additional 3 million cases among children below 15 years of age. Additionally, because people in these areas have not been exposed to malaria before therefore they are particularly vulnerable to succumbing to malaria as it spreads to higher altitudes. The spread of malaria in the highlands of Eastern Africa have also been attributed to land use changes in the form of agroforestry development which is exacerbated by the scarcity of resources (Source). In an earlier study by Pascual and coauthors in 2006 found evidence of increase in temperatures leading to a resurgence in malaria incidence in East African highlands, since the end of the 1970s. Similar migration of malarial parasites and *Ae. aegypti* mosquitoes, which were once limited to warmer temperature, low altitude areas have been found above one mile in the highlands of northern India and at 1.3 miles in the Colombian Andes, since the 1980s. Similar spread of malaria in the highland regions have also been observed in the highland areas of Burundi (Moise et al. 2016).

For instance, in 2013 about 198 million cases, and about 584,000 deaths were reported. Approximately, 90% of these deaths occurred in Africa, and children below the age of 5 years constituted 78% of the malaria deaths worldwide. As shown in Fig. 3.2, there are substantial variations in deaths caused by malaria at the regional scale and different age groups. In Africa and the Eastern Mediterranean region, the number of deaths caused by malaria among females increased from 2000 to 2012. The age specific distribution of deaths caused by malaria exhibited a slight U shaped curve in all regions, with the highest mortality concentrated within the under 5 segment and then a slight increase in the older age groups of 50 years and above. However, the African region experienced a reversal in the declining trend with age in total number of deaths for the oldest age group of 50–59 years. Overall, the Western Pacific region experienced the lowest total number of deaths caused by malaria, which also overlaps with low risk levels for malaria incidence. The African region experienced an increase in the number of deaths caused by malaria in all age groups, which also has one of the highest gender disparities in the Global South. The decline in most of the regions in 2012 compared to 2000 can be attributed to the increased funding for the eradication of malaria across the globe. For instance, over the past 10 years vector control measures such as insecticide-treated mosquito nets were made available to almost half of the population in Sub Saharan Africa, the region most hit by malaria. However, a majority of the countries reporting malaria incidence have experienced insecticide resistance in malaria vectors. Specifically, malaria infection during pregnancy has been identified as a significant risk for pregnant women, her fetus, and the newborn child. This is particularly evident in countries where malaria is endemic, due to physiological and hormonal changes and reduced immunity. Depending on the transmission level, it can lead to spontaneous abortion, stillbirth, premature birth, low birth weights, maternal anemia, and mild to severe malaria. The proportion of women receiving intermittent preventive treatment in pregnancy for malaria has shown an overall increase, however the levels remain below the program targets. This is further evidenced by the higher attendance rates for antenatal care services (WHO 2014). Other than insecticide and drug resistances, issues regarding lack of continued political support and willingness to maintain funding for such eradication programs can also lead to the reemergence of these diseases in some of the countries. This is evidenced by an increase in the number of deaths caused by malaria in the African region. For instance, it is estimated that over

Infectious Diseases

Fig. 3.2 Deaths caused by Malaria among females in 2000 and 2012 (**a**) African Region (**b**) Southeast Asian Region (**c**) Eastern Mediterranean Region (**d**) Western Pacific Region. (Data Source: WHO 2008)

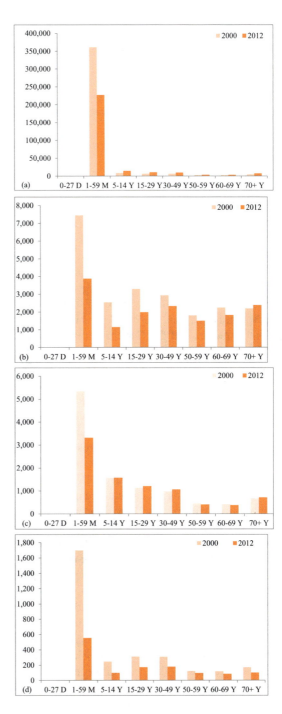

30 million women living in malaria endemic areas of Sub Saharan Africa become pregnant each year(Dellicour et al. 2010), out of which about 10,000 maternal deaths and 75,000–200,000 infant deaths that are caused by malaria (Scott et al. 2014). In addition there is increasing evidence of malarial infection during the first and second year of an infant exposed to blood-stage malarial antigens and gaining a tolerant phenotype that persists in childhood (Malhotra et al. 2009).

Dengue

Dengue fever is a viral infection that is transmitted by bites from infected female Aedes mosquitos. These mosquitos can only thrive in regions where the temperatures rarely fall below 50 °C. The Center for Disease Control (CDC) estimates that more than one third of the world population is at risk for infection. It is one of the leading causes of illness and death in the tropics and subtropics. Approximately 400 million people are infected annually by dengue virus, and it has emerged as a major world-wide problem since the 1950s (CDC 2015). Majority of this disease is concentrated in Southeast Asia and the Western Pacific. In recent years dengue fever has spread in South America and the Caribbean region, with sporadic outbreaks reported from Africa and Middle East. The incidence of dengue fever is estimated to have increased 30-fold over the past 50 years (WHO 2012). This rapid spread of dengue has been attributed to favorable climate conditions related to climate change, urbanization, frequent mobility of people and goods, and lack of trained staff and facilities for proper treatment. Despite its widespread impacts on human health, there is no effective treatment for dengue fever. The incidence of dengue has increased in the recent years in parts of Africa, which is grossly underestimated due to the lack of clinical suspicion and diagnostic tests (WHO/TDR 2009). Furthermore, there has been an increase in dengue outbreaks in South Asia and the Middle East in recent years.

The transmission of dengue occurs through the interaction between humans, mosquitos, viruses, and environmental factors. Higher temperatures lead to increased infection and transmission levels of dengue virus infection rates. Variations in diurnal temperature ranges also affect the infection rates. In addition precipitation influences the habitat availability for mosquito larva and pupae. For example, the interaction between temperature and precipitation influences the rates of evaporation, thus determining the availability of water habitats for mosquito breeding. Precipitation, mainly rainfall is essential to the formation of suitable habitats in the form of stagnant water in containers in urban areas. Several studies have indicated a direct correlation between dengue incidence and the rainy season (Hoeck et al. 2003; Barrera et al. 2011). Additionally, the role of global teleconnections such as La Niña conditions that lead to wetter conditions in Hawaii islands result in the expansion of dengue (Kolivras 2010). On the other hand drier conditions can also lead to increased presence of mosquito breeding habitats, in the form of increased water storages such as domestic water reservoir, water cooler, or large containers used for storing water. Thus it is not only the direct impacts of climate change that

Infectious Diseases

Fig. 3.3 Deaths caused by Dengue among females in 2000 and 2012 (**a**) African Region (**b**) Southeast Asian Region (**c**) Eastern Mediterranean Region (**d**) Western Pacific Region. (Data Source: WHO 2008)

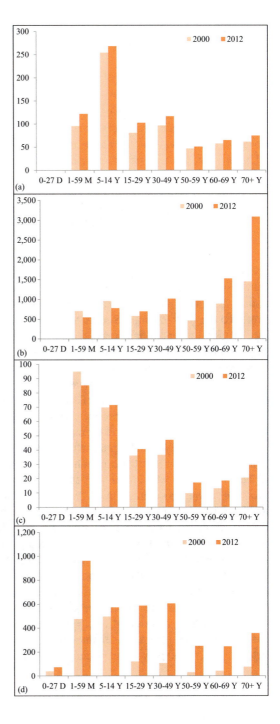

can affect human health but sometimes the adaptive measures people take to mitigate the effect (Morin et al. 2013).

However, the relationship between climate and dengue incidence and transmission is not necessarily linear, thus making it more complex to forecast or prevent. Other than climate variables, the role of vegetation indices, tree cover, housing quality, and surrounding land cover also have affect the incidence patterns (Troyo et al. 2009). For instance, the preference for cooler water temperatures and shaded containers for mosquitos to lay eggs were found in Puerto Rico (Barrera et al. 2006). The opposite, in the form of mosquitos' preference for direct sun exposure was found in Iquitos, Peru (Wong et al. 2011). Furthermore, similar to the malaria vector, greater diurnal temperature range inhibited the exposure of mosquitos to be infected by the dengue virus (Lambrechts et al. 2011).

Overall, dengue incidence worldwide has increased across the Global South, particularly Southeast Asia and the Western Pacific Region where the number of fatalities were greater than 1000 (Fig. 3.3). In these two regions the number of female fatalities from dengue was higher than male fatalities in all age groups. This may be due to the propensity of women to stay at home, taking care of household chores that bring them in contact more often to sources of water, which also are preferred mosquito habitats. Similar to malaria, majority of the fatalities occurred at an early age among children, except in the Southeast Asia region where the majority of the fatalities were concentrated in the older age group of 70 years and greater. Moreover, contrary to malaria, there is a greater spread of dengue related fatalities among different age groups. An increase in dengue related deaths among adolescents and adults, particularly in the Global South has also been noted (Tantawichien 2012). Dengue has a greater presence in different parts of Asia mainly due to more humid conditions accompanied by rapid urbanization. In many of the large metropolitan areas of the Global South, where dengue outbreaks have occurred in recent years, rapid unplanned urbanization has led to improper sewage and waste management. In this context, several studies have indicated the greater vulnerability of pregnant women to dengue fever. For instance in French Guiana, dengue fever among pregnant women led to fetal deaths but no birth defects (Basurko et al. 2009; Carles 1999). In another study examining the incidence of maternal dengue infection in Rio de Janeiro, Brazil showed an increased risk of the development of severe dengue fever among pregnant women and mortality rates (Machado et al. 2013). It is important to note that there are no vaccinations against dengue fever and no clear cure, contributing to potentially a more dangerous disease that has shown a recent reemergence outside its endemic areas. Additionally, recent severe outbreaks of dengue have been reported in large overcrowded metropolitan areas such as New Delhi, India. Therefore, a greater amount of resources need to be devoted toward the detection, cure, and prevention of dengue at various spatial scales.

Extremes

One of the most widespread impacts of projected and already occurring climate change related processes include climate extremes. Extremes refer to extreme climate and extreme weather, often used interchangeably, defined as the occurrence of weather or climate events above or below a certain threshold value. Extreme weather refers to short term extreme and severe weather such as hurricanes, tornados, and flash floods. Extreme climate refers to extreme weather over longer time periods such as heat waves and cold waves. Extreme weather events are considered as a good indicator of long term climate change. It has been repeatedly indicated in numerous scientific studies, that there will be an overall increase in the frequency of climate extremes (Seneviratne et al. 2012). The first decade of the twenty-first century has already been declared as the warmest period on record since modern measurements commenced around 1850 (WMO 2013). This decade was marked by all kinds of record breaking climate extremes, such as record precipitation, 2003 European heat wave, and subsequent heat waves in parts of South Asia, hurricanes including Katrina in the US and Haiyan in Philippines, and 2010 floods in Pakistan. As outlined in Chap. 1, there is high confidence in the significant increase in warm days and nights, and extreme heavy precipitation events in different parts of the world. There is also significant evidence that much of the warming in extreme daily minimum and maximum temperatures has been caused by anthropogenic activities, including GHG emissions and land use land cover changes. There is also medium confidence in the role of anthropogenic activities on the increasing trends in extreme precipitation events. Not all extreme weather is catastrophic, but some of them can lead to disasters and cause widespread damage to human and natural systems. It is important to note that even though there have been increases in the trends of extreme weather events, population has also increased in areas vulnerable to natural disasters, thus increasing the overall impact of such events. Most of this increase in population in vulnerable areas has taken place in the developing world, distributed across the Global South. Additionally, the definition of extreme weather differs spatially depending on the normal. For instance, the definition of a hot day in the tropics is a lot higher temperature compared to that in the mid or high latitudes. Furthermore, the data on extreme weather events are more readily available for developed countries than less developed countries, thus leading to limited analysis for the Global South. Overall, there is general consensus about the greater impacts of extreme weather events on water management systems, agriculture and related food security, health, and tourism. Needless to say, extreme weather events also have a big impact on physical infrastructure, particularly transportation, which in turn can have widespread socio-economic impacts. This section will mainly focus on two aspects of climate extremes which include the impacts of extreme temperature patterns and climate variability, in the form of long term climate processes such as El Niño and Southern Oscillation (ENSO). In subsequent chapters focus on food security and water, the impact of natural disasters such as floods and droughts will be discussed.

Fig. 3.4 Women in South Asia cover their faces to protect themselves from sun burn and air pollution

Heat Waves

One of the most common forms of extreme weather events includes heat waves, used to refer to unusually hot weather over a prolonged period of time. According to the World Meteorological Organization (WMO), an heat wave is defined as a period when the daily maximum temperatures of more than 5 days exceeds the average maximum temperature by 5 °C, with the normal period being 1981–2010. Heat waves are most common in the summer when a high pressure system develops over an area associated with synoptic weather conditions such as the slow movement of an air mass over an area (Met Office 2015). According to a WMO survey, a total of 56 countries reported their highest absolute daily maximum temperature record over the period 1961–2010 during the years 2001–2010. This includes some of most severe heat waves that were observed in 2002 and 2003, followed by the heat wave of 2015 (Fig. 3.4). Each of these heat waves killed more than 1000 people. There were severe heat waves during January to March 2006 in Brazil and pre-monsoon heat wave in Pakistan in 2010 (WMO 2013). The impact of extreme summer heat on human health is aggravated by increases in humidity, which is determined by the heat index.

During heat waves, elderly people and those with pre-existing conditions are most vulnerable. The majority of the excess mortality is caused by cardiovascular, cerebrovascular, and respiratory diseases (McMichael et al. 2003). Other than age, socio-economic status, housing conditions, prevalence of air conditioning have also been cited as factors determining the temperature-related mortality. Additionally, prolonged heat wave like conditions accompanied with reduced night time cooling enhances the development of smog and the dispersal of allergens (Epstein 2000).

Climate Variability

Table 3.1 Relation between ENSO Index and monthly level total dengue cases

Country	El Niño	La Niña	Neutral	Time period
Venezuela	0.68	−0.2633	−0.1789	January 1997 to March 2005
Indonesia	0.02	0.30744	−0.2295	January 1998 to March 2005
Laos	−0.06	−0.0261	0.21169	January 1998 to December 2006
Philippines	−0.32	0.02738	0.15307	January 2006 to August 2010
Singapore	−0.32	0.05552	−0.0652	January 1999 to March 2010
Suriname	−0.09	−0.4562	−0.1468	January 98 to December 2004

Due to limited data availability in the Global South, there are very few extensive studies on the impact of heat waves. However, during the 2003 heat wave in Europe the mortality rates were highest for women and elderly people. Specifically in Portugal it was twice the number of men, in France it was 70% higher than expected for women. In this context, the projected increases in temperatures and frequency of heat waves will aggravate the impacts on women. For instance, in the majority of the rural areas of the Global South, the poorer households do not have piped water supply. In these areas women are responsible for walking long distances to fetch water for the daily needs of their families. The increasing trends in day time temperatures will expose them to more harsh conditions. Poorer women who already have high levels of malnutrition will be easily susceptible to heat strokes and other heat related stresses. In addition, the majority of rural areas, and significant sections of the urban areas in the Global South have poor air conditioning thus making the already vulnerable population susceptible to the variability in weather conditions. In urban areas people are at greater risk to heat related stresses due to urban heat island effects. Another increasingly common feature in some of the larger metropolitan areas in the Global South is growth of slums caused by overcrowding in the cities. Often these slums are located right next to very wealthy areas, and provide a lot of essential services, which are part of informal economy. However, despite their location in these large cities most of the dwellings are temporary and lack proper sanitation.

Climate Variability

Climate variability can be defined as the variation around the long term average climate conditions. It includes seasonal variations and irregular events such as ENSO, which is one of the most dominant modes of climate variability on the interannual time scales. It has widespread impacts on temperatures and rainfall concentrated across the tropics. Due to changes in global circulation patterns, El Niño years are associated with droughts in Asia, Australia, and Northeast Brazil, while floods occurring in parts of South America. There is substantial evidence of the impact of climate variability in the form of ENSO on the incidence of infectious diseases such as dengue and malaria. An El Niño event occurs every 2–7 years (Ebi

Fig. 3.5 Some of the news headlines from major news agencies in India reporting about the recent dengue outbreak in New Delhi and other large cities in India

et al. 2003). For instance, the inter-annual variation in malaria cases in Colombia between 1960 and 1992 showed a close relationship with ENSO, with 17.3% increase in number of malaria cases during an El Niño year followed by a 35.1% decrease in the year after an El Niño event (Bouma et al. 1997). Similar association between El Niño years and malaria incidence was also found in neighboring Venezuela (Bouma and Dye 1997). The role of ENSO are more pronounced in arid areas, where short term rainfall can substantially change the landscape, making it more favorable for the transmission of malaria vectors. For instance, heavy rainfall during La Niña years in the border areas of the Thar Desert in Rajasthan, India resulted in greater incidence of malaria and increased epidemic risk (Akhtar and McMichaels 1996). On the other hand an increase in malaria incidence was also reported in more humid areas as a result of droughts linked to El Niño, when the decreased flow in rivers make it more conducive for mosquito breeding sites (Aron and Patz 2002). In addition, ENSO related droughts can lead to increased mobility among people in search of food and better conditions, thus exposing non immune vulnerable population to the risks of infections (Prothero 1994). An analysis of limited data on total dengue cases at the monthly scale available from WHO (WHO 2015d), showed a clear correlation between the cold, warm, and neutral phases of ENSO for a sample of countries (Table 3.1). For example, there is a strong positive correlation between the total dengue cases during El Niño years and vice versa during La Niña years in Venezuela. However, the relationship is reversed in Philippines, Singapore, and weaker in Indonesia.

Other than global teleconnections like ENSO, seasonal climate variability was found to play an important role in determining the malaria inpatient numbers in several sites in East African Highlands.[1] Furthermore, the variability in climatic conditions may also lead to change in the timing and duration of pollen and spore seasons, thus affecting allergic disorders such as asthma and fever (Beggs 2004). These effects are further aggravated by population increase, unplanned urbanization, land use land cover changes, inadequate health infrastructure, which are typical issues in the Global South. For instance, the recent dengue outbreak in India exposed the weaknesses in the health support infrastructure in the larger cities (Fig. 3.5). This latest dengue outbreak in New Delhi has been identified as the worst in last 5 years and a very rare strain of the virus. This outbreak exposed the weakness in the health care infrastructure here and also the lack of sanitation that caused such a severe outbreak. The outbreak also coincided with the monsoon season, which is the main rainy season of the year across most of the Indian subcontinent. Many of dengue patients were children and elderly, which again bring women to the forefront as main caregivers and thus putting them at risk of infections.

Global Patterns of Female Health Indicators in Relation to Climate Change and Vulnerability Index

The direct and indirect impacts of climate change on human health are also determined by the quality of health care available for local populations. In many of the countries in the Global South, the negative impacts of climate change on human health may be further aggravated by poor quality of health care infrastructure, particularly in the rural areas where the lack of adequate medical supplies and health care support can lead to preventable mortality. The lack of medical facilities is more critical in the rural areas where the poor physical infrastructure results in lack of access to some of the remote areas. In many of the countries in the Global South women and girls may have restricted movement due to local cultural norms. Thus their access to health care is limited. Furthermore, poor and inadequate health care is particularly detrimental for pregnant women who cannot travel long distances for health care. In addition, women are often the primary caregivers at home bearing the bulk of the responsibility of all household chores, including cooking, collecting water, and taking care of the unwell. They are also often forced to stay close to their communities, in sometimes unhygienic conditions leading to increase in gynecological problems (WHO 2015e). In majority of the Global South, the lack of decision making power among women also poses serious threat to their health. Furthermore, the findings of the Lancet Commission of Women and Health emphasized the double burden of communicable and non-communicable diseases faced by women in many of the low income countries in the Global South (Langer et al. 2015).

In Fig. 3.6, the amount of out-of-pocket expenditures out of total health care expenditures and the satisfaction levels of the quality of available health care have

---

[1] Zhou et al. (2004).

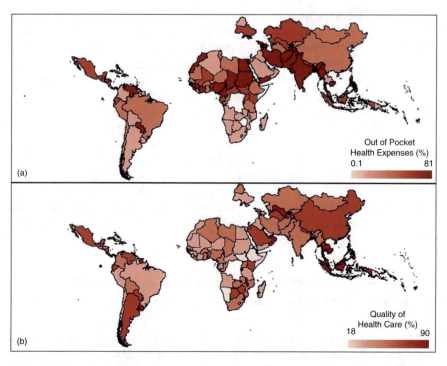

Fig. 3.6 Spatial patterns of (**a**) Out-of-pocket expenses out of total health expenditure 2008–2012; (**b**) Quality of health care satisfaction level in 2011

been mapped. In view of the relatively scarce gender disaggregated data available on health indicators, these two variables provide valuable insight into the level of health care disparities in the Global South. Interestingly, most of South Asia, which consistently showed the highest level of gender inequality, has relatively higher levels of satisfaction with regard to the quality of health care available. However, most of South America and Africa showed substantially lower levels of satisfaction with the quality of health care available. The lower levels of satisfaction about existing health care in these countries may also be attributed to the high rates of migration of healthcare workers from Sub Saharan Africa, resulting in a deficit of 2.4 million doctors and nurses (Naicker et al. 2009). This is particularly relevant in the context of one of the specific targets of the third SDG goal, which aims to increase the financing and recruitment, retention of health workforce in developing countries and small island states like Philippines. Women in most of these countries do not have access to basic health services and modern contraceptives. Specifically, the burden of communicable diseases and perinatal and nutritional disorders remain very high in Sub Saharan Africa (Gates 2015). However, the out of pocket expenses toward total health care expenses were substantially high across most of Sahel region of Africa, South and Southeast Asia, and Brazil and Peru in South America. Higher out of pocket expenses can be disadvantageous to women and girls' health, due to lower levels of empowerment and decision making power and the lower preference for the girl child. This is also important for maternal health and child care in the Global South. In a recent report released by the McKinsey Global Institute, found the unmet need for family planning needs among 197 million women globally, who want to stop or delay having children (McKinsey Global Institute 2015). The quality of health care is not only for women and girls health, but also for the overall welfare of the society, which is a direct indicator of gender parity. Women as the primary caregivers are therefore affected the most by these services. Finally, a healthier adult woman raising a family who is aware of health and wellbeing will lead to healthier adults in the future and better care giving. In a recently published study shows women's contributions in the health system was 2.35% of the global Gross Domestic Product (GDP) for unpaid work and 2.47% of GDP for paid work, which equates to about $3.052 trillion (Langer et al. 2015).

Therefore, it is important to develop gender sensitive and gender responsive policies. However, it is important to mention that gender sensitive policies do not only mean "adding on" a concern for women, but also incorporate deeper understanding of gendered forms of vulnerability. It also should include a greater commitment of resources, financial, technical, and human, to address specific gendered priorities (Ahmed and Fajber 2009). Therefore, gender sensitive assessments will lead to gender responsive interventions and policies, which will ultimately lead to more effective climate change mitigation and adaptation at all levels. It is important to take into consideration the incidence of major infectious diseases based on gender because of the different gendered roles that creates a differentiated impact on men and women.

References

Ahmed, S., & Fajber, E. (2009). Engendering adaptation to climate variability in Gujarat, India. *Gender and Development, 17*, 33–50.

Akhtar, R., & McMichael, A. J. (1996). Rainfall and malaria outbreaks in western Rajasthan. *Lancet, 348*, 1457–1458.

Alonso, D., Bouma, M. J., & Pascual, M. (2011). Epidemic malaria and warmer temperatures in recent decades in an East African highland. *Proceedings of the Royal Society B: Biological Sciences, 278*, 1661–1669.

Aron, J., & Patz, J. A. (Eds.). (2002). *Ecosystem change and public health*. Baltimore: Johns Hopkins University Press.

Bacaër, N., & Guernaoui, S. (2006). The epidemic threshold of vector borne diseases with seasonality. *Journal of Mathematical Biology, 53*, 421–436.

Barrera, R., Amador, M., & Clark, G. G. (2006). Ecological factors influencing Aedes aegypti (Diptera: Culicidae) productivity in artificial containers in Salinas, Puerto Rico. *Journal of Medical Entomology, 43*, 484–492.

Barrera, R., Amador, M., & MacKay, A. J. (2011). Population dynamics of Aedes aegypti and dengue as influenced by weather and human behavior in San Juan, Puerto Rico. *PLoS Neglected Tropical Disease, 5*. https://doi.org/10.1371/journal.pntd.0001378.

Basurko, C., Carles, G., Youssef, M., & Guindi, W. E. (2009). Maternal and foetal consequences of dengue fever during pregnancy. *European Journal of Obstetrics & Gynecology and Reproductive Biology, 147*, 29–32.

Beggs, P. (2004). Impact of climate change on aeroallergens: Past and future. *Clinical and Experimental Allergy, 34*, 1507–1513.

Béguin, A., Hales, S., Rocklöv, J., Åström, C., Louis, V. R., & Sauerborn, R. (2011). The opposing effects of climate change and socio-economic development on the global distribution of malaria. *Global Environmental Change, 21*(4), 1209–1214.

Bouma, M. J., & Dye, C. (1997). Cycles of malaria associated with El Niño in Venezuela. *The Journal of the American Medical Association, 278*, 1772–1774.

Bouma, M. J., Poveda, G., Rojas, W., Chavasse, D., Quinones, M., Cox, J., et al. (1997). Predicting high-risk years for malaria in Colombia using parameters of El Niño Southern Oscillation. *Tropical Medicine & International Health, 2*, 1122–1127.

Caminade, C., Kovats, S., Rocklov, J., Tompkins, A. M., Morse, A. P., Colón-González, F. J., et al. (2014). Impact of climate change on global malaria distribution. *Proceedings of the National Academy of Sciences., 111*, 3286–3291.

Campbell-Lendrum, D., Manga, L., Bagayoko, M., & Sommerfeld, J. (2015). Climate change and vector-borne diseases: What are the implications for public health research and policy? *Philosophical Transactions of the Royal Society of London. Series B, Biological Sciences, 370*, 20130552.

Carles, G. (1999). Effects of dengue fever during pregnancy in French Guiana. *Clinical Infectious Diseases, 28*, 637–640.

CDC. (2015). *Where malaria occurs*. Retrieved September 9, 2015, from http://www.cdc.gov/malaria/about/distribution.html

Chaves, L. F., & Koenraadt, C. J. M. (2010). Climate change and highland malaria: Fresh air for a hot debate. *The Quarterly Review of Biology, 85*(1), 27–55.

Dellicour, S., Tatem, A. J., Guerra, C. A., Snow, R. W., & ter Kuile, F. O. (2010). Quantifying the number of pregnancies at risk of malaria in 2007: A demographic study. *PLoS Medicine, 7*, e.1000221.

Ebi, K. L., Mearns, L. O., & Nyenzi, B. (2003). Weather and climate: Changing human exposures. In A. J. McMichael, D. H. Campbell-Lendrum, C. F. Corvalán, K. L. Ebi, A. K. Githeko, J. D. Scheraga, et al. (Eds.), *Climate change and human health: Risks and responses*. Geneva: World Health Organization.

Emert, V., Fink, A. H., & Paeth, H. (2013). The potential effects of climate change on malaria transmission in Africa using bias-corrected regionalised climate projections and a simple malaria seasonality model. *Climatic Change, 120*, 741–754.

Epstein, P. R. (2000). Is global warming harmful to health? *Scientific American, 283*, 50–57.

Gates, M. (2015). Valuing the health and contribution of women is central to global development. *The Lancet, 386*. https://doi.org/10.1016/S0140-6736(15)60940-0.

Gillies, M. T. (1953). The duration of the gonotrophic cycle in Anopheles gambiae and An. funestus with a note on the efficiency of hand catching. *East African Medical Journal, 30*(1953), 129–135.

Githeko, A. K., Lindsay, S. W., Confalonieri, U. E., & Patz, J. A. (2000). Climate change and vector-borne diseases: A regional analysis. *Bulletin of the World Health Organization, 78*, 1136–1147.

Hoberg, E. P. (2015). Evolution in action: Climate change, biodiversity dynamics and emerging infectious disease. *Philosophical Transactions of the Royal Society B: Biological Sciences, 370*. https://doi.org/10.1098/rstb.2013.0553.

Hoeck, P. A. E., Ramberg, F. B., Merrill, S. A., Moll, C., & Hagedorn, H. H. (2003). Population and parity levels of Aedes aegypti collected in Tucson. *Journal of Vector Ecology, 28*, 65–73.

Horton, R., & Ceschia, A. (2015). Making women count. *The Lancet, 386*, 1112–1114.

Koenraadt, C. J. M., Githeko, A. K., & Takken, W. (2014). The effects of rainfall and evapotranspiration on the temporal dynamics of Anopheles gambiae ss and Anopheles arabiensis in a Kenyan village. *Acta Tropica, 90*(2), 141–153.

Kolivras, K. N. (2010). Changes in dengue risk potential in Hawaii, USA, due to climate variability and change. *Climate Research, 42*, 1–11.

References

Lambrechts, L., Paaijmans, K. P., Fansiri, T., Carrington, L. B., Kramer, L. D., Thomas, M. B., et al. (2011). Impact of daily temperature fluctuations on dengue virus transmission by Aedes aegypti. *Proceedings of the National Academy of Sciences, 108*, 7460–7465.

Langer, A., Meleis, A., Knaul, F. M., Atun, R., Aran, M., Arreola-Ornelas, H., et al. (2015). Women and health: The key for sustainable development. *The Lancet, 386*, 1165–1210. https://doi.org/10.1016/S0140-6736(15)60497-4.

Lozano, R., Naghavi, M., Foreman, K., Lim, S., Shibuya, K., Aboyans, V., et al. (2012). Global and regional mortality from 235 causes of death for 20 age groups in 1990 and 2010: A systematic analysis for the global burden of disease study 2010. *The Lancet, 380*, 2095–2128.

Machado, C. R., Machado, E. S., Rohloff, R. D., Azevedo, M., Campos, D. P., de Oliveira, R. B., et al. (2013). Is pregnancy associated with severe dengue? A review of data from the Rio de Janeiro surveillance information system. *PLoS Neglected Tropical Diseases, 7*. https://doi.org/10.1371/journal.pntd.0002217.

Malhotra, I., Dent, A., Mungai, P., Wamachi, A., Ouma, J. H., Narum, D. L., et al. (2009). Can prenatal malaria exposure produce an immune tolerant phenotype? A prospective birth cohort study in Kenya. *PLoS Medicine, 6*, e1000116.

Martens, W. J. M., Niessen, L. W., Rotmans, J., Jetten, T. H., & McMichael, A. J. (1995). Potential impact of global climate change on malaria risk. *Environment Health Perspectives, 103*, 458–464.

Martens, W. J. M., Jetten, T. H., Rotmans, J., & Niessen, L. W. (1995). Climate change and vector-borne diseases: A global modelling perspective. *Global Environment Change, 5*, 195–209.

Martin, P. H., & Lefebvre, M. G. (1995). Malaria and climate: Sensitivity of malaria potential transmission to climate. *Ambio, 24*, 200–207.

Matsuoka, Y., & Kai, K. (1994). An estimation of climatic change effects on malaria. *Journal of Global Environmental Engineering, 1*, 1–15.

McGeehin, M. A., & Mirabelli, M. (2001). The potential impacts of climate variability and change on temperature-related morbidity and mortality in the United States. *Environment Health Perspectives, 109*, 185–189.

McKinsey Global Institute. (2015). *The power of parity: How advancing women's equality can add $12 trillion dollars.* Retrieved October 2, 2015, from www.mckinsey.com/mgi

McMichael, A. J., Wooodruff, R. E., & Hales, S. (2006). Climate change and human health: Present and future risks. *The Lancet, 367*, 859–869.

McMichael, A. J., Campbell-Lendrum, D. H., Corvalán, C. F., Ebi, K. L., Githeko, A., Scheraga, J. D., et al. (2003). *Climate change and human health: Risks and responses.* Geneva: World Health Organization.

Met Office. (2015). *Heat wave.* Retrieved September 17, 2015, from http://www.metoffice.gov.uk/learning/learn-about-the-weather/weather-phenomena/heatwave

Moise, Imelda K., et al. (2016). Seasonal and geographic variation of pediatric malaria in Burundi: 2011 to 2012. *International journal of environmental research and public health 13*(4), 425.

Morin, C. W., Comrie, A. C., & Ernst, K. (2013). Climate and dengue transmission: Evidence and implications. *Environmental Health Perspectives, 121*, 1264–1272.

Naicker, S., Plange-Rhule, J., Tutt, R. C., & Eastwood, J. B. (2009). Shortage of healthcare workers in developing countries—Africa. *Ethnicity and Diseases, 19*, S1–S60.

Paaijmans, K. P., Read, A. F., & Thomas, M. B. (2009). Understanding the link between malaria risk and climate. *Proceedings of the National Academy of Sciences, 106*, 13844–13849.

Patz, J. A., Graczyk, T. K., Geller, N., & Vittor, A. Y. (2000). Effects of environmental change on emerging parasitic diseases. *International Journal for Parasitology, 30*(12), 1395–1405.

Prothero, R. M. (1994). Forced movements of population and health hazards in tropical Africa. *International Journal Epidemiology, 23*, 657–664.

Rogers, D. J., & Randolph, S. E. (2000). The global spread of malaria in a future, warmer world. *Science, 289*, 1763–1766.

Rueda, L. M., Patel, K. J., Axtell, R. C., & Stinner, R. E. (1990). Temperature-dependent development and survival rates of Culex quinquefasciatus and Aedes aegypti (Diptera: Culicidae). *Journal of Medical Entomology, 27*, 892–898.

Scott, S., Mens, P. F., Tinto, H., Nahum, A., Ruizendaal, E., Pagnoni, F., et al. (2014). Community-based scheduled screening and treatment of malaria in pregnancy for improved maternal and infant health in The Gambia, Burkina Faso and Benin: Study protocol for a randomized controlled trial. *Trials, 15*, 340.

Seneviratne, S. I., Nicholls, N., Easterling, D., Goodess, C. M., Kanae, S., Kossin, J., et al. (2012). Changes in climate extremes and their impacts on the natural physical environment. In *Managing the risks of extreme events and disasters to advance climate change adaptation, A special report of working groups I and II of the Intergovernmental Panel on Climate Change* (p. 109). Cambridge: Cambridge University Press.

Tantawichien, T. (2012). Dengue fever and dengue haemorrhagic fever in adolescents and adults. *Pediatrics and International Child Health, 32*, 22–27.

Troyo, A., Fuller, D. O., Calderón-Arguedas, O., Solano, M. E., & Beier, J. C. (2009). Urban structure and dengue fever in Puntarenas, Costa Rica. *Singapore Journal of Tropical Geography, 30*, 265–282.

Turell, M. J. (1989). Effects of environmental temperature on the vector competence of Aedes fowleri for Rift Valley fever virus. *Research in Virology, 140*, 147–154.

UN. (2016). SDGs: Sustainable development knowledge platform. *United Nations.* Retrieved December 29, 2016, from https://sustainabledevelopment.un.org/sdgs

Watts, D. M., Burke, D. S., Harrison, B. A., Whitmire, R. E., & Nisalak, A. (1987). Effect of temperature on the vector efficiency of Aedes aegypti for dengue 2 virus. *American Journal of Tropical Medicine and Hygiene, 36*, 143–152.

WHO. (2008). *The global burden of disease: 2004 update.* Geneva: Author.

WHO. (2012). *Global strategy for dengue prevention and control.* Geneva: Author.

WHO. (2014). *World malaria report 2014.* Geneva: Author.

WHO. (2015a). *The world health report 2002: Reducing risks promoting healthy life.* Retrieved August 10, 2015, from http://www.who.int/whr/2002/en/whr02_en.pdf?ua=1

WHO. (2015b). *Climate change and health fact sheet n°266.* Retrieved August 10, 2015, from http://www.who.int/mediacentre/factsheets/fs266/en/

WHO. (2015c). *Global health estimates 2014 summary tables: Deaths by cause, age and sex, by WHO region, 2000-2012.* Retrieved September 29, 20015, from http://www.who.int/mediacentre/factsheets/fs310/en/index4.html

WHO. (2015d). *DengueNet database and geographic information system.* Retrieved September 30, 2015, from http://apps.who.int/globalatlas/DataQuery/default.asp

WHO. (2015e). *Gender, climate, change and health.* Retrieved October 1, 2015, from http://apps.who.int/iris/bitstream/10665/144781/1/9789241508186_eng.pdf?ua=1v

WHO. (2017). *Climate change and health fact sheet n°266.* Retrieved November 14, 2017, from http://www.who.int/mediacentre/factsheets/fs266/en/

WHO and WMO. (2015). *Atlas of health and climate.* Retrieved August 10, 2015, from http://www.who.int/globalchange/publications/atlas/report/en/

WHO/TDR. (2009). *Dengue guidelines for diagnosis, treatment, prevention and control.* Geneva: WHO.

WMO. (2013). *The global climate 2001–2010: A decade of climate extremes summary report.* Geneva: Author.

Wong, J., Stoddard, S. T., Astete, H., Morrison, A. C., & Scott, T. W. (2011). Oviposition site selection by the dengue vector Aedes aegypti and its implications for dengue control. *PLoS Neglected Tropical Disease, 5*. https://doi.org/10.1371/journal.pntd.0001015.

Zhou, G., Minakawa, N., Githeko, A. K., & Yan, G. (2004). Association between climate variability and malaria epidemics in the East African highlands. *Proceedings of the National Academy of Sciences, 101*, 2375–2380.

Chapter 4
Water

Water is the driving force of all nature. Leonardo da Vinci
Sustainable Development Goal 6: "Ensure availability and sustainable management of water and sanitation for all" (UN 2016).

Introduction

In January 2015, heavy seasonal rainfall across Southern Africa resulted in flooding Malawi, Mozambique, Madagascar, and Zimbabwe. Around 135,000 people were affected by these floods, with many left homeless. Many people are still trying get back to normal with their daily activities. By March, as a result of torrential rains, floods devastated some of the direst parts of Peru and Chile. More than 100,000 people were left without electricity over extended periods of time in Chile. This was followed by one of the most severe heat waves in three decades in May and June 2015, across South Asia. It resulted in 1000 deaths in Pakistan and more than 2000 deaths in India. These conditions were not only because of higher temperatures but also a result of lower air pressure and high humidity in some areas. This was followed by extreme heavy rains associated with the annual monsoon season in South Asia and Cyclone Komen that made landfall in Bangladesh. This is despite the lower than normal monsoon rainfall forecasted for this year by the India Meteorological Department. Heavy rains caused the derailment of two trains in Central India killing more than a dozen passengers. Monsoon related fatalities have been reported in Pakistan (206), India (166), and Myanmar (96). In addition, people have been forced to evacuate as a result of flooding in various parts of South Asia, which includes 803,271 poeple in Pakistan and 330,000 in Myanmar, who have been forced to evacuate or affected by floods. All of these natural disasters highlight the vulnerability of the Global South to the higher frequency of occurrence of extreme weather events along with the lower levels of preparedness in terms of infrastructure and people living in these extremely vulnerable areas.

© Springer International Publishing AG, part of Springer Nature 2018
S. Sen Roy, *Linking Gender to Climate Change Impacts in the Global South*,
Springer Climate, https://doi.org/10.1007/978-3-319-75777-3_4

In November 2015, South Africa was declared to be experiencing the worst drought since 1982. This drought has affected more than 2.7 million households, approximately 18% of the population, with water shortages, thus affecting their livelihoods and draining the economy. The most severely affected states include Free State and KwaZulu-Natal (Al Jazeera 2015). Droughts were also reported in 2015 in the South Pacific island of Vanuatu, which was hit by category five storm, Cyclone Pam earlier in the year. The drought was mainly triggered by a strong El Niño in the Pacific Ocean. It was reported that the droughts have affected mostly the local subsistence crops of fruits and vegetables, leading to people surviving on the government supplied tinned fish and rice (Voice of America 2015).

The above mentioned events are all related to either water scarcity or surplus in the same year in different locations across the Global South. These events characterize natural disasters occurring during a typical year in the context of changing climate. The abundance or scarcity of water affects not only in the form of droughts and floods, but also availability of basic needs such as food and shelter, as well as health and infrastructure.

In this regard, it is important to examine some of the basic facts about water availability and consumption across the earth's surface and changes in the contest of climate change. Two thirds of the earth's surface is covered with water, of which less than 3% is fresh water. Most of this fresh water is inaccessible, because only 0.3% of the freshwater is available from streams, rivers, and lakes (National Geographic 2015). The majority of the freshwater is captured in glaciers and ice caps. Clean water is an essential resource for human health, vibrant and diverse ecosystem, and economic growth. Water is an integral part of the climate system in the form of the hydrological cycle, gaseous, liquid, or solid form, and therefore plays a major role in the projected changes in climate system. For instance, climate models project an increase in levels of water vapor in the atmosphere as a result of increased evapotranspiration related to increase in near surface air temperatures. This in turn will lead to warmer air temperatures, because water vapor is a well-known GHG absorbing outgoing radiation from the earth surface. Additionally, increased evapotranspiration from the earth surface will lead to lower soil moisture, ground water storage, and stream flow. Due to our incomplete understanding of the interaction of water vapor-cloud-climate, there are substantial differences in the model outputs. One of the reasons for our limited understanding is the scarcity of continued long term measurements of hydrological characteristics on the earth surface such as ground water and soil moisture (NOAA 2015). The importance of water in the climate system is not only in the atmosphere, but it also plays an important role through the oceans by transferring heat from the lower latitudes to higher latitudes.

However, the availability of water is one of the main manifestations of climate change on the earth's surface, through the timing and distribution and thus affecting the livelihood and well-being of societies. For instance, higher temperatures and changes in the frequency and intensity of extreme weather events are predicted to affect the availability and distribution of precipitation, snowmelt, stream flow, groundwater storage, water quality, and sea levels. These changes will have the

Introduction

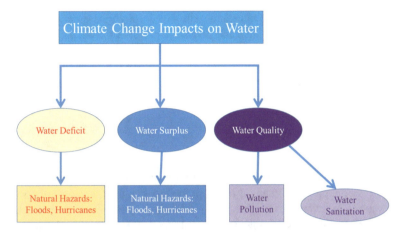

Fig. 4.1 Brief summary of the impacts of climate change related processes on water resources

greatest impacts on the poorer and vulnerable societies in the Global South, who are not adequately prepared for such changes. According to the World Bank, currently 1.6 billion people are living in the absolute water scarcity, which is expected to rise to 2.8 billion people by 2025 (World Bank 2015). This is particularly critical in view of the projected population increase from 7.3 billion now to 8.5 billion by 2030 and 11.2 billion in 2100 (United Nations, Department of Economic and Social Affairs, Population Division 2015), which will be mostly concentrated in the Global South.

Specifically, it is estimated that the majority of the population growth will be concentrated in Africa, which currently has the highest population growth rates of 2.55% annually between 2010 and 2015. The latest estimates indicate that approximately half of the world's population growth between 2015 and 2050 will be concentrated in nine countries: India, Nigeria, Pakistan, Democratic Republic of Congo, Ethiopia, Tanzania, U.S.A., Indonesia, and Uganda. Other than the U.S.A., the rest of all the countries are located in the Global South with some of the highest gender gaps. These countries are also very vulnerable to the impacts of climate change, such as droughts, floods, and sea level rise. This projected steep increase in populations in the Global South will present challenges for the already stressed existing health and education support systems, along with unstable political systems in some of the countries. Therefore, Africa will soon have the majority of the proportion of younger population in the world. On the other hand, there are also projections of a significant growth in aging populations across South and Central America and Asia, where the proportion of aging population will increase from the current levels of 11–12% of the population above the age of 60 years to 25% by 2050. This will put additional stress on the health care system and food supplies. This will in turn put immense pressure on the finite resources on the earth surface, in the form of freshwater availability for daily activities and irrigation for agricultural purposes. In general, there will be an unprecedented increase in the demand for freshwater resources. Therefore, it is critical to address the impacts of climate change directly on the availability of freshwater resources and indirectly through extreme weather events such as floods, food security, and sea level rise. The rest of the chapter will examine the impact of climate change related processes on water resources with regards to issues related to water surplus or scarcity or quality (Fig. 4.1), with specific focus on how they impact women and girls.

Water Scarcity

One of the direct effects of climate change on water resources will be in the form of reduced water supply in already drought prone areas in the Global South. Scarcity of the water supply on the earth's surface not only affects the overall water supply for daily needs but also soil moisture available for plants, and groundwater supply. Water scarcity is directly related to the hydrological cycle as well as land use land cover changes. In view of the rapid rates of urbanization accompanied by the burgeoning population in the Global South, there will be increased demand for water. In the Middle East and North Africa (MENA) region the groundwater utilization exceeds the average annual recharge by a factor of three times or more (Alavian et al. 2009).

Some of the key projections regarding the changes in the hydrological cycle pertinent to water scarcity issues include increased rates of evapotranspiration and reduction in soil moisture, and changes in precipitation regime. Decreasing trends in precipitation the form of reduced amount and frequency of rainfall are predicted in the southern edge of Amazonia and South and Central Brazil (Cook et al. 2012), and form of precipitation like reduced snowfall in high altitude areas such as the Himalayas (Basannagari and Kala 2013). Reduced snowfall is already causing decreased flow in some of the major rivers of South Asia, which are lifelines of water for the agriculture intensive areas located near the rivers. Additionally, model projections indicate a decline in precipitation and increase in the amount of evapotranspiration in the already dry regions of MENA (Terink et al. 2013). Overall, there is an increasing trend in the length and intensity of occurrence of droughts in the tropics and subtropics (Trenberth et al. 2014). Furthermore, some of the currently observed changes in the hydrological cycle are projected to intensify in the future, such as the increasing trends in droughts in the tropics and subtropics. In general most of the extremely high and high water stress areas are concentrated in the densely populated regions of the Global South, including South and East Asia, and MENA regions of the world. There is relatively less water stress in South and Central America, except parts of Chile and Argentina and eastern Brazil. In a lot of water stressed regions of the world, surface water shortage is replaced by groundwater retrieval. According to the World Bank, over half of the world's population depends on groundwater for everyday uses. Specifically in the developing world approximately 20–40% of total water use is fulfilled by groundwater.

A recent study using satellite data to assess the depletion of groundwater in Northwestern India showed a rate of 1 foot per year decline in groundwater levels over the past decade. Most of the groundwater has been pumped out for irrigation and other daily activities (Rodell et al. 2009). Historical analysis indicates a significant increase in the frequency and spatial spread of droughts particularly in the tropics since the 1970s, which is more likely caused by human activities. This increasing trend in droughts has been mainly caused by decreasing precipitation and increased temperature, resulting in increased evapotranspiration and reduced soil moisture (Alavian et al. 2009). Furthermore, the projected trends in droughts also show an increasing trend by the 2090s, particularly in regions that are heavily dependent on glacial flows (Bates et al. 2008). Overall, majority of the projected water stressed areas are mostly clustered in the Global South.

Thus, climate change will lead to water insecurity in parts of Asia, which have experienced water security historically, and it will worsen water security issues in MENA where it is already a scarce resource. Furthermore, the decrease in water availability will also make international water sharing disputes more controversial, such as in the case of the Mekong River shared by five countries in East and Southeast Asia. It is estimated that currently 1.6 billion people live in countries and regions experiencing water scarcity, which is projected to rise to 2.8 billion people by 2025. With increasing population and decline in evapotranspiration rates in the arid semi-arid regions of the world, there will be greater demand for water for irrigation in agricultural areas essential for food security. This will result in an increase in the potential dangers of the occurrence of conflicts in regions where there is water sharing disputes, particularly in water scarce regions in the Global South (Wolf et al. 2003). In a later chapter focusing specifically on the impact of climate change on conflicts and its subsequent burden on women and girls, this will be examined in greater detail.

One of the most significant direct impacts of water scarcity will be on women and girls. In most of the rural areas of the Global South, women and girls are responsible for water collection for daily needs such as cooking and other household activities. In view of projected increase in the length and intensity of droughts combined with the drying up of local sources of water, women and girls will have to walk longer distances (sometimes 30 km) (Develop Africa 2015), to collect water on their backs or heads for their daily requirements. This is further aggravated by harsher weather conditions in the form of higher temperatures and unfavorable land cover (such as lesser number of shade providing trees), which will make them more vulnerable to heat strokes and dehydration. In many cases walking longer distances in search of water takes them on unexplored and unsafe paths, thus exposing them to not only the vagaries of nature but also physical and emotional harassment from problem elements. Reports of harassments of women and girls walking off the beaten track paths are regularly reported all across the Global South. Often girls at a very young age are exposed to such harassment.

Other than walking longer distances to fetch water, carrying water in pots and other storage containers over long distances can pose a health hazard in the future on their lower backs, spines, and neck muscles. In some cases it may lead to early ageing or the vertebral column (Mehretu, and Mutambirwa 1992a). With increasing temperatures, there will increased need for per capita water consumption, which may require multiple trips and long waits at the sources of water. This excessive burden on women and children also affect the nutritional needs and health maintenance. The excessive time devoted to the domestic chores by women and girls in Sub Saharan Africa in terms of long distances is well documented (Mehretu, and Mutambirwa 1992b), which will further increase with predicted changes in climate in the future. Additionally, the analysis of data collected through qualitative and quantitative surveys in South Africa showed that most of the water collection chores are done by girls at a young age (Hemson 2007). As a result of the extended times spent on collecting water during the day by young girls, many times they are late to school or miss school, resulting in lower levels of self-confidence. Lower education levels will ultimately result in them being less competitive in the labor market, and lower levels of awareness about health and control over their personal lives. Additionally, in a recent study focusing on water needs and women's health in Ghana showed that in traditional settings, under water scarcity conditions, women would often give priority to their husbands' water needs over their own (Buor 2003).

Additionally, in many of the countries in the Global South, when there are droughts girls are held back at home to help out in the household chores and take care of their younger siblings while the parents go out to work. Thus, younger girls are not able to attend school or get the proper education from a very young age. Furthermore, in many cases in the rural areas when there is a drought men leave the villages in search of better opportunities to support their families back home. This result in women as the main caregiver at home left under very difficult conditions of food and water scarcity. Often in these situations women are in a very vulnerable position in the hands of moneylenders with illegal terms and conditions. They also experience limited freedom due to the predominantly patriarchal societies in most of the regions. Moreover, in some cases women and girls have to engage in unfavorable occupations to support themselves and their families, including prostitution which exposes them to life threatening diseases such as HIV and other STDs.

In some of the mountainous villages of Bolivia, as a result of shrinking of glaciers, women have to walk longer distances in treacherous mountainous terrains to collect water (UN Population Fund 2009). Similarly, the recent droughts during summer 2015 in Cambodia exposed women in their primary role as homemakers to harsher weather conditions. Water scarcity also meant that they could not plant their staple crops such as the water intensive rice and other vegetables. This is particularly critical for the subsistence and rain fed agriculture economies in the Global South, which are mainly dependent on rain water for irrigation. It is also estimated based on available data, that the quantity of water collected per capita is reduced significantly if the walk to the source of water takes 30 min or longer (WHO and UNICEF 2006).

It is estimated that by cutting 15 min off the walking time to a source of water could help in the reduction of child mortality less than five years by 11% and nutrition depleting diarrhea by 41%. Specifically, in Ghana, a 15 min reduction in the walking time for water collection led to an increase from 5 to 12% in girls' school attendance (Nauges and Strand 2011). However, it is noteworthy in the context of droughts in the Global South where it has led to suicides by male farmers after major crop failures as a result of droughts in the semi-arid regions of India in view of the lost income and mental stress (Behere and Behere 2008).

Water Surplus

One of the frequently observed aspects of variability in weather associated with climate change is extreme precipitation events. These events constitute the occurrence of very heavy rainfall within short intervals of time that can cause substantial damage to life and property. These can cause substantial damage to property and human life. As mentioned in previous chapters, most analysis of these extreme precipitation events indicates a significantly positive trend over an extended period of time (Sen Roy and Balling 2004; Wang and Zhou 2005). Such events lead to floods and inundation of the existing infrastructure, which can cause severe limitations in terms of mobility and accessibility to essential services. There are increasing frequencies of 100 year floods that now take place every 10 or 20 years. This increasing trend has been attributed to climate change in the form of radiative effects of anthropogenic changes in atmospheric composition, which leads to the intensification of global water cycle (Cubasch 2001). This results in increased flood risk. In a detailed analysis of discharge observations across 29 large river basins, revealed substantial increase in the frequency of great floods during the twentieth century (Milly 2002). This is further aggravated by the results of ensemble model projections, which reveal an increase in the global exposure to floods based on the degree of warming. At the regional scale, some of the greatest increases are projected in South and Southeast Asia, Eastern Africa, and northern half of the Andes (Hirabayashi 2013). The frequency of large floods are further aggravated by global teleconnections such as ENSO, which is considered as one of the most dominant inter-annual teleconnections impacting global scale climate processes (Ward et al. 2014).

Floods are both the most common natural disaster and number one cause of mortality globally between 2005 and 2014 (IFRC 2016). Additionally, floods and extreme temperatures caused the maximum average annual mortality, while floods and droughts accounted for the highest number of people affected by natural disasters globally between 2005 and 2014. During the same period, Asia and Africa reported the highest number of floods related disasters. More specifically, the highest number of people killed by floods was reported in Asia, while the number of people killed due to droughts was highest in Africa (Fig. 4.2). Floods are projected to increase manifold in the future as well as population living in vulnerable areas. It is also noteworthy, that mortality resulting from extreme temperatures was highest in Europe and high human development countries. It is evident from the 2003 heat wave in Europe, which resulted in more than 70,000 fatalities (Robine et al. 2008). France was one of the worst hit countries with 14,802 deaths. A significant number of people affected were elderly population of 65 years and above. Moreover, the number of female deaths was significantly higher in France and Europe. However,

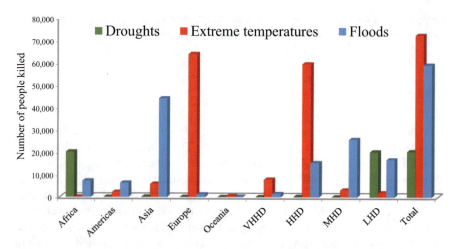

Fig. 4.2 Total number of people reported killed, by type of phenomenon, continent and level of human development (2005–2014). (*VHHD* Very High Human Development, *HHD* High Human Development, *MHD* Medium Human Development, and *LHD* Low Human Development according to UNDP human Development Index) (Data Source: IFRC 2016)

the vulnerability to climate change from hydrological extremes in the tropics is critical due to frequency and lower levels of preparedness of people living here.

In this context, the effect of natural disasters such as floods are not gender neutral, rather there is increasing evidence of the greater burden of the effects of climate change on women and girls, predominantly in the Global South. One of the main causes of the disproportionate burden of natural disasters like floods is existing social norms, which inhibit women from having access to local resources and capacity to respond with resilience. Despite limited gender disaggregated data available on impact of natural disasters on women, several specific natural disasters-related case studies reveal the greater burden on women. For instance the effect of the 2004 Indian Ocean Tsunami resulted in 13,000 fatalities in Sri Lanka, of which 65% were females (Oxfam International 2005). Most of the females were in the age group of 19–29 years. The number of female deaths outnumbered males in the other countries affected by the 2004 Tsunami such as India and Indonesia. The higher mortality of females compared to males during the tsunami has been attributed to lower levels of awareness among women and girls, lower access to information, and the lack of skills to survive rising waters including swimming (Oxfam International 2005). Additionally, it was surmised that weaker physical strength and lower stamina among women and girls in Indonesia made them more vulnerable to the Tsunami, and cultural and social norms restricted women from leaving their homes during floods in Bangladesh (Rohr 2006). Furthermore, in many of these countries the traditional clothing women wear make them less mobile during a natural disaster. Similarly, in 2008 cyclone Nargis in Myanmar resulted in 85,000 fatalities, of which 61% were women and girls (Myanmar Government and Association of Southeast Asian Nations and United Nations 2008).

Floods are not only prevalent in Asia, but it impacts large parts of Africa also. For instance two thirds of Nigeria's states are under serious threats of floods, such as in 2012. Additionally, the densely populated capital city of Lagos is not only under serious threat of floods due to lack of adequate drainage facilities, but also sea level rise (Ugwu and Ugwu 2013). In addition to the greater physical vulnerability of women and children to floods, there is also evidence of greater emotional trauma caused by floods on female victims. Particularly, life in relief camps after floods can be especially trying on women, who have to fend for the elderly and children of the family. They also become more exposed to harassment and lesser personal security. Globally 15 countries, all of which are located in the Global South, account for 80% of populations exposed to flood risks worldwide (World Resources Institute 2015). It is also noteworthy that a majority of these 15 countries also have some of the highest observed gender gaps. Finally, due to greater proportion of women living in poverty, they are economically disadvantaged when faced with the repercussions of a natural disaster.

Water Quality

It is predicted that the projected changes in precipitation and temperature regimes will have a direct impact on not only water availability, but also the water quality. The impacts will be on both surface and ground water. It is estimated that by 2050, about 50% of the total population will be living in countries experiencing water stress related to climate change (IFPRI 2012). Additionally, global water demand is projected to increase by 55% by 2050, with most of the increase concentrated in emerging economies and developing countries located in the Global South that are already under immense water stress (OECD 2012; Wang et al. 2013). Already 2.4 billion people (36% of the world population approximately) are experiencing water stress (IFPRI 2012). It is widely established that climate change will alter the hydrological cycle through reduced or excessive precipitation in different regions of the world. Specifically, evapotranspiration is predicted to increase in most areas as a result of increased temperatures. Water stress is also projected to increase in view of population increase, urbanization, and economic development in already water stressed areas of the world (Blignaut and van Heerden 2009; Wang et al. 2013). Some of the indirect impacts of climate change on water availability for daily uses, also referred to as domestic water demand, has a direct impact on human health and living conditions. Several regional level studies have been conducted to assess the impact of climate change on domestic water use demand. For instance, an increase of 0.3% of domestic water use was predicted in Thailand (Jampanil et al. 2012) but no significant change in Tehran, Iran (Karamouz et al. 2011).

Additionally, the impacts of climate change will also be felt on the overall water quality that can be measured in the form of access to drinking water and sanitation, and the spread of water borne diseases. The latest available data from WHO/ UNICEF Joint Monitoring Programmed for Water Supply and Sanitation 2015 on

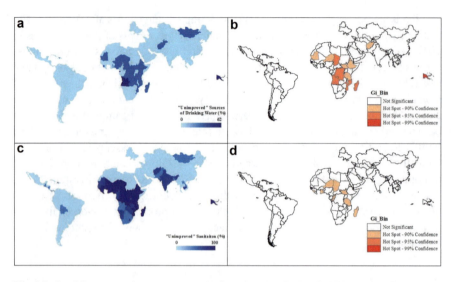

Fig. 4.3 Spatial patterns of access to sources of water and sanitation facilities in the Global South in 2015 (**a**) Access to unimproved sources of drinking water; (**b**) Hot spots of unimproved sources of drinking water; (**c**) Access to unimproved sources of sanitation; (**d**) Hot spots of unimproved sources of sanitation. (Data Source: WHO and UNICEF 2016)

the access to unimproved sources of drinking water and unimproved forms of sanitation at the national level have been mapped (Fig. 4.3). An unimproved source of drinking water is defined as a source that does not adequately protect from outside contamination, particularly fecal matter. Therefore the typical unimproved drinking water sources include unprotected springs, unprotected dug wells, carts with small trunks/drums, tanker-trucks, surface water, and bottled water (WHO and UNICEF 2016). Unprotected springs and dug wells are subject to runoff, bird droppings, or the entry of animals. Water from tanker-trucks or carts with trunk or drum, which consists of water transported to communities in tankers or trunks or drum, can be subject to contamination due to unhygienic conditions and improper cleaning or unimproved sources of water. Bottled water is considered to be unimproved on a case by case basis when the overall source of drinking water is not an improved source of water. Countries where greater than 25% of the population did not have access to improved sources of drinking water are mainly concentrated in Sub Saharan Africa, including Congo DRC, Nigeria, Ethiopia, Kenya, and Tanzania, and semi-arid regions in Central Asia, including Afghanistan, Tajikistan, and Mongolia. In Angola and Equatorial Guinea, more than 50% of the population did not have access to improved sources of drinking water, while it was 60.3% of the population in Papua New Guinea. All of these countries also rank high in terms of gender inequality indices, which makes women and girls more prone to the ill effects of accessing drinking water from unimproved sources of water. Additionally, none of the countries in South and Central America and the Caribbean ranked highly in terms of access to unimproved sources of drinking water. It is noteworthy that 2.1

billion people have gained access to improved sources of drinking water since 1990, most of which has occurred in the Global South. However, only 84% of the global rural population had access to improved sources of drinking water, compared to 96% of global urban population. At the regional scale, in Sub Saharan Africa, 29% of the population (37% in rural areas and 14% in urban areas) was reported to be at least 30 min away from an improved source of drinking water.

In addition to access to drinking water, it is also important to have proper sanitation in place for healthy living conditions. As evident from Fig. 4.3, the percentage of population living in unimproved sources of sanitation conditions are much higher and widespread in the Global South. Unimproved sources of sanitation include no facilities (aside from bush or field), shared sanitation facilities, including hanging toilet, bucket or other container for retention of feces; pit latrines; and pour flush facilities where excreta is deposited nearby the household environment. Poor sanitation conditions can potentially be a major environmental hazard for people living under such situations. It is also more concerning that densely populated regions in South and Southeast Asia, such as India, had more than 50% of their population living under unhygienic sanitation conditions. In Sub Saharan Africa, countries with greater proportion of population with access to unimproved sources of water overlapped with more than 50% of their population living with poor levels of sanitation. Furthermore, almost half of the global rural population did not have access to improved levels of sanitation, which were higher in the Global South. Shared sanitation facilities were more widespread in urban areas than in the rural areas where open defecation and other forms of unimproved sanitation are more prevalent. It is noteworthy that the urban population has increased by 73%, compared to 11% in the rural areas since 1990. Future population projections indicate a more rapid rate of increase in urban population in the developing and less developed countries of the world, with limited expansion in existing infrastructure and facilities. Sub Saharan African countries are hotspots for both access to unimproved sources of drinking water and sanitation. It is estimated that about half of the total 946 million people who defecate in the open are women.

It is important to note that without good hygiene the full benefits of improved sources of drinking water and sanitation cannot be realized. Personal hygiene and sanitation is particularly critical for health, safety, and dignity of girls and women, particularly menstrual hygiene management (MHM). Though there are limited data available on this issue, it is estimated that about 500 million women and girls lack adequate MHM facilities (WHO and UNICEF 2016). Other than health hazards, basic minimum sanitation is essential for healthy living conditions. Several different case studies have shown that gender sensitive sanitation conditions in schools can result in fewer gender gaps in school attendance and educational attainment, commonly referred to as WASH (water sanitation hygiene) (Mahon and Fernandes 2010; Sommer 2010). In view of long term predictions of increased frequency and prolonged periods of droughts, water scarcity will only make improved sanitation conditions only difficult to achieve. It would lead to more contamination of sources of drinking water, which can have fatal impacts on human health. In this context, given the role of women in collecting water for the household, they are more prone to the detrimental impacts of water contamination and poor sanitary conditions resulting from water scarcity. Thus it is imperative to incorporate gender sensitive sanitation measures at the grassroots level. The negative consequences from poor unimproved sources of drinking water and inappropriate sanitation measures result in a non-ending vicious cycle, unless adequate measures are taken at an earlier stage.

Waterborne Diseases

In view of the critical role of the availability of safe drinking water and clean sanitation on good health, this final section focuses on the impact of water scarcity or surplus on the health of women and girls in the Global South. As mentioned above, most of the people with limited access to improved sources of drinking water are located in rural areas. However, it is important to note that water contamination and limited access to piped water supply still continues to be a major issue in many of the larger metropolitan areas in the Global South. For instance, in New Delhi, it is common for a majority of the residents to be without water for days during the hot summer months, and with limited hours of water supply during other months. In addition, places that have improved sources of water, such as piped water supplies, may get contaminated due to improper maintenance. Moreover, during major natural disasters like hurricanes, water contamination is a major problem. When the source of water is far away, then the quantity and quality of water being compromised are high. Thus in many cases the water needs of the household may not be adequately fulfilled, particularly for the female members of the house who have lesser priority in the household hierarchy.

This limited supply of water also leads to the degradation of poor hygiene and sanitation practices, which in turn can lead to the spread of infectious diseases and affect overall health. Additionally, a wide assessment of health care centers in low and middle income countries indicated that a majority of these centers lacked basic water, sanitation, and hand washing facilities. Most of these centers lacking basic facilities were located in Africa, Latin America, South and Southeast Asia, and the Caribbean. Specifically in Africa, about 42% of the facilities lacked in improved source of water within 500m. The result of poor sanitation is directly translated to spread of infections among patients during a hospital stay. The incidences of infections among newborns are widely prevalent, with sepsis and other severe infections causing 430,000 deaths annually. Poor sanitation can also endanger pre and postnatal care for women (WHO and UNICEF 2015). In general, early child-bearing along high fertility rates increase the risk of poor health and lower levels of education, which further result in lower income levels and empowerment (WHO 2016). Lower levels of empowerment can result in unwanted or unplanned pregnancies, which put additional pressure on the limited available resources.

It is estimated that about 842,000 people died due to diarrhea in 2012, resulting from inadequate improved sources of drinking water, sanitation, and hygiene in low and middle income countries. Most of these deaths were concentrated again in Africa (44%) and South and South East Asia (43%). A gender disaggregated total indicated that females accounted for 49% of these deaths in Africa and 59% in South and Southeast Asia. Additionally, the female death rates were higher in all age groups in Asia. This gender disparity in the form of higher number of deaths among females can be attributed to the patriarchal norms of the society, which puts women and girls at a disadvantage in terms of accessing proper health care (WHO and UNICEF 2015). The situation was more glaring for children in India and Bangladesh, where the hospitalization rates were slower or delayed in cases of girls versus boys (Malhotra and Upadhyay 2013). Similar health care disparities were also reported in Sub Saharan Africa, where the percentage of children with diarrhea who did not receive medical advice was higher for girls than boys (Kanamori and Pullum 2013).

The prevailing social norms can be particularly detrimental for women's health during natural disasters such as hurricanes and floods. For instance, during the 1998 floods of Bangladesh, there was an increase in the incidence of urinary tract infections and rashes among young girls due to poor sanitation and hygiene. Similarly, during the 1991 Bangladesh floods, more women and children died because they were waiting for their husbands to return and make the decision to leave (WHO 2017a). In many cases, when there is a major natural disaster, then most of the affected people, mainly poorer people, are evacuated to relief camps. Many of these relief camps have inadequate sanitation and medical facilities. This makes it particularly difficult for those who are unwell to recover. Furthermore, as a result of a lack of security for girls and makeshift toilet facilities render women and girls especially vulnerable. In many cases, women and girls are responsible for most of the household chores; therefore they are also at high risk of exposure to contaminated water during washing and cleaning household items.

One of the major impacts of natural disasters such as floods and hurricanes is the spread of infectious diseases such as cholera and diarrhea. Given the already increasing trends in extreme precipitation events and associated flooding, there is increased risk of water borne diseases. As shown in Fig. 4.4, some of the more densely populated tropical countries in the Global South, including Nigeria, India, Democratic Republic of Congo (DRC), have a large total number of cholera cases. However, some of the African countries experienced a large number of cases in the recent decades (2010 and 2012), which can be directly attributed to water contamination and unimproved sources of water during flooding events, in addition to poor sanitation. The latest available estimates for DRC during 2015 indicate 19,705 cases of cholera, with some of the largest numbers reported from South Kivu (4906), ex-Katanga (4565), Maniema (3971), and North Kivu (3294) provinces (WHO 2017b). This is particularly critical, in view of some of the provinces such as South Kivu, which has many refugee camps with refugees from Burundi. In many of the affected areas there is limited coverage of improved sources of water, with the some of the water around the main lakes in this region being areas of multiplication of cholera-related bacteria. Due to poor infrastructure and a rapidly increasing population, most of Sub-Saharan Africa is predominantly prone to cholera epidemics as observed during 2011 and 2015.

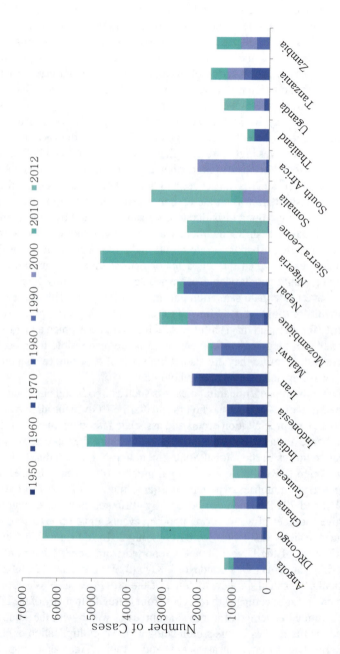

Fig. 4.4 Total Number of Cholera cases in selected countries of the Global South

References

Al Jazeera. (2015). South Africa in midst of 'epic drought'. *AlJazeera.com*. Retrieved November 4, 2015, from http://www.aljazeera.com/news/2015/11/south-africa-midst-epic-drought-151104070934236.html

Alavian, V., Qaddumi, H. M., Dickson, E., Diez, S. M., Danilenko, A. V., Hirji, R. F., et al. (2009). *Water and climate change: Understanding the risks and making climate-smart investment decisions*. Last updated November 2009, from http://www-wds.worldbank.org/external/default/WDSContentServer/WDSP/IB/2010/02/01/000333038_20100201020244/Rendered/PDF/529110NWP0Box31ge0web0large01128110.pdf

Basannagari, B., & Kala, C. P. (2013). Climate change and apple farming in Indian Himalayas: A study of local perceptions and responses. *PLoS One, 810*, e77976. https://doi.org/10.1371/journal.pone.0077976.

Bates, B. C., Kundzewicz, Z. W., Wu, S., & Palutikof, J. P. (Eds.). (2008). *Climate change and water*. Technical Paper of the Intergovernmental Panel on Climate Change, Geneva.

Behere, P. B., & Behere, A. P. (2008). Farmers' suicide in Vidarbha region of Maharashtra state: A myth or reality? *Indian Journal of Psychiatry, 50*, 124–127.

Blignaut, J., & van Heerden, J. (2009). The impact of water scarcity on economic development initiatives. *Water SA, 35*, 415–420.

Buor, D. (2003). Water needs and women's health in the Kumasi Metropolitan Area, Ghana. *Health and Place, 10*, 85–103.

Cook, B., Zeng, N., & Yoon, J.-H. (2012). Will Amazonia dry out? Magnitude and causes of change from IPCC climate model projections. *Earth Interactions, 16*, 1–27. https://doi.org/10.1175/2011EI398.1.

Cubasch, U. (2001). Projections of future climate change. In J. T. Houghton, Y. Ding, D. J. Griggs, M. Noguer, P. J. van der Linden, X. Dai, et al. (Eds.), *Climate change 2001: The scientific basis*. Cambridge: Cambridge University Press.

Develop Africa. (2015). *The effect of drought on women and children*. Retrieved November 12, 2015, from https://www.developafrica.org/blog/effect-drought-women-and-children

Hemson, D. (2007). The Toughest of Chores': Policy and practice in children collecting water in South Africa. *Policy Futures in Education, 5*. https://doi.org/10.2304/pfie.2007.5.3.315.

Hirabayashi, Y. (2013). Global flood risk under climate change. *Nature, 3*, 816–821.

IFPRI. (2012). Chap. 3: Sustainable food security under land, water, and energy stresses. In *Global hunger index*. Washington, DC: Author.

IFRC. (2016). *IFRC World Disasters Report: Focus on local actors, the key to humanitarian effectiveness*. Retrieved December 30, 2016, from http://ifrc-media.org/interactive/world-disasters-report-2015/

Jampanil, D., Suttinon, P., Seigo, N., & Koontanakulvong, S. (2012). Application of input-output table for future water resources management under policy and climate change in Thailand: Rayong province case study. *Change, 27*, 29.

Kanamori, M. J., & Pullum, T. (2013). Indicators of child deprivation in sub-Saharan Africa: Levels and trends. In *Demographic and health surveys, DHS comparative reports 32*. Calverton, MD: ICF International.

Karamouz, M., Zahmatkesh, Z., Nazif, S. (2011). Selecting a domestic water demand prediction model for climate change studies. In Proceedings of the 2011 world environmental and water resources congress: Bearing knowledge for sustainability, California.

Mahon, T., & Fernandes, M. (2010). Menstrual hygiene in South Asia: A neglected issue for WASH (water, sanitation and hygiene) programmes. *Gender & Development, 18*, 99–113.

Malhotra, N., & Upadhyay, R. P. (2013). Why are there delays in seeking treatment for childhood diarrhea in India? *Acta Paediatrica, 102*, e413–e418. https://doi.org/10.1111/apa.12304.

Mehretu, A., & Mutambirwa, C. (1992a). Gender differences in time and energy costs of distance for regular domestic chores in rural Zimbabwe: A case study in the Chiduku communal area. *World Development, 20*, 1675–1683.

Mehretu, A., & Mutambirwa, C. (1992b). Time and energy costs of distance in rural life space of Zimbabwe: Case study in the Chiduku Communal Area. *Social Science and Medicine, 34*, 17–24.

Milly, P. C. D. (2002). Increasing risk of great floods in a changing climate. *Nature, 415*, 514–517.

Myanmar Government and Association of Southeast Asian Nations and United Nations. (2008). *Post-Nargis recovery and preparedness plan*. Retrieved November 14, 2017, from http://myanmar.unfpa.org/sites/default/files/pub-pdf/PONREPP.pdf

National Geographic. (2015). *Earth's freshwater*. Retrieved November 5, 2015, from http://education.nationalgeographic.com/media/earths-fresh-water/

Nauges, C., & Strand, J. (2011). *Water hauling and girls' school attendance: Some new evidence from Ghana, Policy research working paper No. 6443*. Washington, DC: The World Bank.

NOAA. (2015). *Water and climate*. Retrieved November 4, 2015, from http://www.esrl.noaa.gov/research/themes/water/

OECD. (2012). *The OECD environmental outlook to 2050: The consequences of inaction*. The Hague: OECD and the PBL Netherlands Environmental Assessment Agency.

Oxfam International. (2005). The tsunami's impact on women. *Oxfam Briefing Note, March 2005*. Retrieved November 14, 2017, from https://www.oxfam.org.nz/sites/default/files/reports/The_tsunami_impact_on_women.pdf

Robine, J. M., Cheung, S. L. K., Le Roy, S., Van Oyen, H., Griffiths, C., Michel, J. P., et al. (2008). Death toll exceeded 70,000 in Europe during the summer of 2003. *Comptes Rendus Biologies, 331*, 171–178.

Rodell, M., Velicogna, I., & Famiglietti, J. S. (2009). Satellite-based estimates of groundwater depletion in India. *Nature, 460*, 999–1002.

Rohr, U. (2006). Gender and climate change. *Tiempo, 59*, 3–7.

Sen Roy, S., & Balling, R. C. (2004). Trends in extreme daily precipitation indices in India. *International Journal of Climatology, 24*, 457–466.

Sommer, M. (2010). Where the education system and women's bodies collide: The social and health impact of girls' experiences of menstruation and schooling in Tanzania. *Journal of Adolescence, 33*, 521–529.

Terink, W., Immerzeel, W. W., & Droogers, P. (2013). Climate change projections of precipitation and reference evapotranspiration for the Middle East and Northern Africa until 2050. *International Journal of Climatology, 33*, 3055–3072.

Trenberth, K. E., Dai, A., van der Schrier, G., Jones, P. D., Barichivich, J., & Briffa, K. R. (2014). Global warming and changes in drought. *Nature Climate Change, 4*, 17–22.

Ugwu, L. I., & Ugwu, D. I. (2013). Gender, floods and mental health: the way forward. *International Journal of Asian Social Science, 3*, 1030–1042.

UN. (2016). SDGs: Sustainable development knowledge platform. *United Nations*. Retrieved December 29, 2016, from https://sustainabledevelopment.un.org/sdgs

UN Population Fund. (2009). *State of the world population: Women, population and climate*. New York: Author.

United Nations, Department of Economic and Social Affairs, Population Division. (2015). *World population prospects: The 2015 revision, key findings and advance tables*. Working Paper No. ESA/P/WP.241.

Voice of America. (2015). *Drought compounds hardship on cyclone-hit vanuatu*. Retrieved November 4, 2015, from http://www.voanews.com/content/drought-compounds-hardship-on-cyclone-hit-vanuatu/3036154.html

Wang, X.-J., Zhang, J. Y., Yang, Z. F., Shahid, S., He, R. M., Xia, X. H., et al. (2013). Historic water consumptions and future management strategies for Haihe River basin of Northern China. *Mitigation and Adaptation Strategies for Global Change*. https://doi.org/10.1007/s11027-013-9496-5.

Wang, Y., & Zhou, L. (2005). Observed trends in extreme precipitation events in China during 1961–2001 and the associated changes in large-scale circulation. *Geophysical Research Letters, 32*. https://doi.org/10.1029/2005GL022574.

References

Ward, P., Jongman, B., Kummu, M., Dettinger, M. D., Weiland, F. C. S., & Winsemius, H. C. (2014). Strong influence of El Niño Southern Oscillation on flood risk around the world. *Proceedings of the National Academy of Sciences, 111*, 15659–15664.

WHO. (2016). *Gender, climate change and health*. Retrieved December 30, 2016, from http://apps.who.int/iris/bitstream/10665/144781/1/9789241508186_eng.pdf

WHO. (2017a). *Gender and health in disasters*. Retrieved November 14, 2017, from http://www.who.int/gender-equity-rights/knowledge/a85575/en/

WHO. (2017b). *Cholera – Democratic Republic of the Congo*. Retrieved November 14, 2017, from http://www.who.int/csr/don/15-december-2015-cholera-drc/en/

WHO and UNICEF. (2006). *Meeting the MDG drinking water and sanitation target: The urban and rural challenge of the decade*. Geneva: Author.

WHO and UNICEF. (2015). *Water, sanitation and hygiene in health care facilities: Status in low- and middle-income countries and way forward*. Geneva: WHO.

WHO and UNICEF. (2016). *Joint Monitoring Programme (JMP) for water supply and sanitation*. Retrieved December 30, 2016, from http://www.wssinfo.org/

Wolf, A. T., Stahl, K., & Macomber, M. F. (2003). Conflict and cooperation within international river basins: The importance of institutional capacity. *Water Resources Update, 125*: Universities Council on Water Resources.

World Bank. (2015). *Water and climate change*. Retrieved November 5, 2015, from http://water.worldbank.org/topics/water-resources-management/water-and-climate-change

World Resources Institute. (2015). *World's 15 countries with the most people exposed to river floods*. Retrieved December 30, 2016, from http://www.wri.org/blog/2015/03/world%E2%80%99s-15-countries-most-people-exposed-river-floods

Chapter 5
Climate Refugees

Climate change can enhance the competition for resources—water, food, grazing lands—and that competition can trigger conflict.—Antonio Guterres, the U.N. High Commissioner for Refugees.

As armed conflict is highly gendered and women's experiences during war differ from those of men, any conflict mapping and tracking exercise undertaken for use in negotiation must also take account of shifting gender relations and women's activities throughout all phases of conflict and its aftermath.—Dave Barry

Introduction

According to the latest report of the IPCC, in view of the uncertainties associated with future climate conditions and projected increases in the frequency of extreme weather events, there is medium evidence and high agreement about the increased displacement of people (IPCC 2014). Most of this displacement will be concentrated in the low income densely-populated low lying areas of the Global South. People who are forced to move out of their homes due to stress caused by climate change and global warming are called climate refugees, who are a subset of environmental refugees. Some of the direct impacts of climate change that lead to forced displacement of population include droughts leading to food insecurity, and sea level rise leading to loss of livelihood and homes. Indirect impacts would include conflicts as result of food insecurity particularly in Africa. One of the main aspects of climate change is global warming in the form of rising temperatures. The effect of rising temperatures is visible in the rapid rate of melting of ice caps in the high altitude and latitude areas, which leads to sea level rise and ultimate flooding of densely-populated low lying areas, particularly in the tropics. Rising temperatures also cause droughts and desertification, and conversion of arable agricultural land to deserts, thus making them uninhabitable. In addition, the impacts of climate change on the natural ecosystems, such as fisheries can increase rivalry between fishing nations or transboundary sharing of water resources. These changes can be sudden in the form of extreme weather events like hurricanes and floods or gradual over time like desertification. Specifically, gradual environmental changes such as desertification, coastal and soil erosion, which attract less urgent attention, can have much more long-lasting impacts on human population. Irrespective of the time scale, these events affect the poorest sections of the society most disproportionately, the majority of whom are women and girls. Some of the specific events include the Pakistan floods in 2010, which displaced 1.8 million people and destroyed 1.6 million homes, Hurricane Haiyan displaced more than 4 million people in Philippines.

The displacement of vulnerable population from familiar surroundings has far-reaching impacts in the form of limited access to employment opportunities, education, health, and empowerment. These impacts are greater on women and girls in most of the Global South due to the lower status of women and girls as shown in the form of gender gap and inequalities in previous chapters. According to the UNHCR, 31.7 million people in 2012 and 22 million people in 2013 were displaced by natural disasters, most were from the least developed countries in the Global South. Specifically in 2013, 14.2 million people were forced to leave their homes as a result of storms. The projected number of refugees in 2050 is up to 200 million people, which is a widely quoted estimate by Dr. Norman Myers of Oxford University and the Stern Review Report (Conisbee and Simms 2003). It literally means, that 1 in every 45 people in the world will be displaced by climate change. Due to the definition and circumstances under which climate and environmental refugees are displaced, they are not protected by international laws. This is particularly concerning due to the extenuating circumstances resulting from climate change under which people have to leave their homes. Despite limited data and statistics available about migration triggered by climate change and environmental disasters, some of the important statistics about climate refugees are outlined below:

- On average, about 27 million people are displaced due to climate and weather related disasters each year (Myers 2016; Stern 2007).
- Over the last six years, approximately 2% of the global population has been displaced due to climate and weather-related disasters (Myers 2016; Stern 2007).
- It is estimated that as a result of the rise in global temperatures, there will be a 56% increase in the frequency of intergroup conflicts across world (Hsiang et al. 2013).
- The impact of climate change on possible security implications requiring serious attention among the international community was tabled successfully by an alliance of small island nations in the UN General Assembly resolution 63/281 in 2009 (UNGA 2016).
- Additionally, climate change has been identified as not necessarily the main reason but an important decisive factor in triggering conflicts in situations where there are pre-existing stressors. This can be in the form of access to natural resources, which are already spread very thin and uneven, depletion resulting in conflicts and violence.

Introduction 95

- In the case of small island nations, their very existence and sovereignty are threatened by the impacts of climate change.
- The threat of climate change on small low-lying island nations or resulting conflicts are not only limited to the affected nations, but they also spill over the borders in neighboring countries.
- Post-conflict nations are also extremely fragile and susceptible to the impacts of climate change (EJF 2014).
- Among the world's 19 megacities, 16 are located on coastlines, and the majority of these are in the Global South.
- Finally, in the last six years it is estimated about 1 person was displaced by a climate or weather-related disaster every second (EJF 2014).

However, despite the growing numbers and awareness about climate refugees they are not recognized by international law. According to the current definition of refugees by the UNHCR, it only covers those people who are trying to flee their homes because of state-led persecution based on race, religion, political opinion, or ethnicity. Climate refugees are instead considered to be Internally Displaced Persons (IDPs), because it is estimated that most of the displacement will occur within national boundaries. This may not hold true in all cases such as the impacts of sea level rise in the low lying Pacific Islands, or conflicts arising out of food insecurity caused by long persistent droughts. Overall, labeling them as IDPs gives them fewer rights to cross international borders. The use of the term "refugees" has been considered controversial, with a lot of opposition from many of the developed countries. This opposition is due to the fact that it forces them to offer them the same rights as refugees. However, in view of the projected number of climate refugees in the near future, the wide ranging impacts of climate refugees for the origin and destination locations cannot be ignored. Other alternative terms used to describe climate refugees include *climate evacuee* (implies temporary movement within national borders), *climate migrant* (emphasizes more on the "pull" factor of the destination than the "push" factor of the origin), and *forced migrant* (defined as the non-voluntary displacement of population due to growing impacts of climate change) (Brown 2016).

It can also be said that migrations, both temporary and permanent, are the most effective and quickest methods of dealing with the adverse impacts of climate change. Additionally, as migrants or displaced people or evacuees, women and girls are more vulnerable than men in terms of security and health. In this context, the Nansen Initiative is an important step toward a state led consultative process in creating awareness about the rising impacts of climate change in displacing people from their homes, particularly across international borders. This initiative was endorsed by 110 countries, with one of the key principles being the critical role of states in the prevention and preparation of increased displacement of people in the future with proper effective policies. Specifically, these policies should not discriminate and be based on empowerment, participation, or partnerships. It also emphasized that the greatest attention be given to the most vulnerable, to prevent a

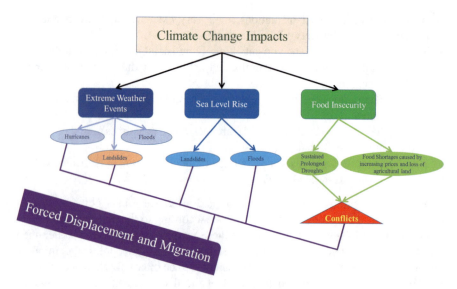

Fig. 5.1 Forced Displacement and migration of population as a result of processes related to climate change

humanitarian crisis and help build resilience to the impacts of climate change (The Nansen Initiative 2016).

According to the UNHCR, women and girls comprise half of any refugee or displaced persons population. Moreover, out of a total of 26 million people displaced by impacts of climate change, approximately 20 million were women mostly in the Global South (WECAN 2016). However, due to lower levels of security and less rights, they are often subject to sexual and gender based violence. The cultural norms in many of the societies in the Global South puts women and girls in more risk due to fewer rights and gendered roles in the society, thus making them more vulnerable. The impacts of climate change driving forced displacement of population from their homes can be classified under three main categories, which include extreme weather events (such as hurricanes, floods), sea level rise, and food insecurity resulting from severe and prolonged droughts (Fig. 5.1). Some of these impacts, particularly food has led to conflicts, such as observed in the case of Syria and Darfur, Sudan. In the majority of the cases, the forced displacement of population is internal within the national borders. However, in many cases it has led to the displacement and out migration of people across international borders. This has resulted in an unprecedented crisis in the receiving countries, which are not prepared to accommodate this influx of population. However, this displacement of vulnerable population from their familiar surroundings puts them at very high risk in terms of security, health, and exploitation. Therefore, the focus of this chapter will be on impact of climate change induced migration and displacement of population, and their impacts on women and girls through specific case studies and analyses of limited available empirical data.

Sea Level Rise

One of the major impacts of climate change is the melting of glaciers and ice caps leading to sea level rise. The causes of sea level rise are not only limited to rise in overall levels of coastal water, but in some cases can be a result of subsidence due to sediment compaction and loading in river deltas, as observed in the case of the Ganges-Brahmaputra delta in Bangladesh. The subsidence in major deltas is caused by the extraction of oil and natural gas, groundwater, and the consolidation from the construction of roads and buildings. Some of the estimated subsidence during the last 100 years in coastal areas in some of the major cities in the Global South includes ~4.4 m in eastern Tokyo, ~2.6 m in Shanghai, and ~1.6 m in Bangkok (Syvitski 2009). In addition to the long-term sustained increases in sea levels across the globe, there are also short-term increases in sea level that can cause irreversible damages to the local coastal communities. These include storm surges resulting from severe storm events like tropical cyclones, heavy rainfall events, and king tides. The impacts of sea level rise are aggravated by the increased rate of urbanization and industrialization in some of the large coastal cities, which increases the costs of the impacts of sea level rise on the communities. Furthermore, fluctuations in global climate in the form of teleconnections such as El Niño can have significant short-term impacts on the variability in sea levels in the range of ±20 to 30 cm (Becker et al. 2012).

> **Box 5.1 Subsiding Coastal Megacities**
> Many of the large densely populated metropolitan cities in the tropics are located in low lying coastal areas, thus making them more vulnerable to the impacts of sea level rise. In a recently published study, 20 cities most at risk of flooding from sea level rise were identified, most of which are located in the Global South. Some of these cities include Khulna, Bangladesh, Kolkata and Mumbai, India, Guangzhou and Shenzen, China, Osak-Kobe, Japan, and Jakarta, Indonesia (Hallegate et al. 2013). These cities are not only vulnerable to the impacts of rising sea levels but also are experiencing land subsidence at the same time. The land subsidence in some cases is occurring at a faster rate than sea level rise. For instance, the densely populated North Jakarta has already sunk 4 m in the last 35 years, which is approximately about 10–20 cm per year (Erkens 2016). Some of the main causes of subsidence are uncontrolled groundwater extraction along with rapid rates of urbanization, in the form of sediment compaction due to building of roads, dredging of harbors, and changes in sediment supply cause erosion. Subsidence rates can also be caused by oil extraction. Specifically, over the last 100 years subsidence of ~4.4 m in eastern Tokyo, ~3 m in the Po delta, ~2.6 m in Shanghai, and ~1.6 m in Bangkok have been observed (Teatini et al. 2011). Furthermore, natural processes such as earthquakes can also lead vertical movement of land as

experienced in the 2011 Japan earthquake that caused a subsidence of 1.2m off the Pacific coast, and excessive sediment compaction and loading caused by some of the major rivers like the Ganges and Brahmaputra in Bangladesh (Syvitski 2009). Coastal subsidence will also lead to increased rates of beach erosion, more persistent flooding, loss of wetlands, increased salinization of aquifer and groundwater, which would have significant impacts on the local infrastructure and food security, human health.

The impacts of sea level rise are already being felt tremendously by coastal communities located in low-lying small islands as densely-populated coastal areas in the Global South. In addition to sea level rise, low-lying coastal communities are also vulnerable to inundation resulting from storm surge and coastal flooding, which have become more frequent and can cause lasting damage to infrastructure. According to the latest report of the IPCC, the population and assets exposed to the detrimental impacts of sea level rise will significantly increase in the future due to population growth, economic development, and urbanization (Wong et al. 2014). It is estimated that the global mean sea levels have risen by 0.19 (0.17–021) m during 1971–2010, based on tidal gauge records and satellite measurements (Rhein et al. 2013). Due to greater confidence in rising sea levels, coastal systems and low lying areas will increasingly be faced with the impacts of sea level rise and losing their homes. It is significant in view of the fact that 44% of the population live within 150 km of the coast (Laczko and Aghazarm 2009). Spatially, the regions that will be most affected by the effects of sea level rise are located in the densely-populated coastlines and islands of East, Southeast, and South Asia (Fig. 5.2a). The amount of land lost due to sea level rise in 2010 is highest in the densely populated island nations of Southeast Asia, particularly Indonesia, Philippines, which is projected to double by 2030 This loss of land is related to the sinking of most of the large deltas due to groundwater withdrawal, floodplain engineering, and trapping sediments by dams. It is estimated that the projected rise in sea level will affect about a million people in South and Southeast Asia (Hijioka et al. 2014). The amount of land lost due to sea level rise is also higher in Central America and the Caribbean (Fig. 5.2b). There is high agreement about the benefits of protection against coastal flooding and loss of land as a result of sea level rise, despite limited evidence.

At the regional scale, the small island developing states (SIDS) in the tropics are most vulnerable to the impacts of sustained sea level rise, caused mainly by melting glaciers and ice caps, storm surge from tropical cyclones, and the effects of global teleconnections like El Niño. Most of the small island nations are located in the Pacific Ocean, Caribbean Sea, Indian Ocean, and the Eastern Atlantic. The effects of sea level rise will not only affect the physical infrastructure, but also the local economy which is highly dependent on the oceans in the form of tourism and marine life. It will also result in an increase in the levels of salinity in the soil, rendering it unusable for agriculture. Rise in the salinity levels in fresh water bodies will also impact fisheries and water supply to the local communities, who are heavily depen-

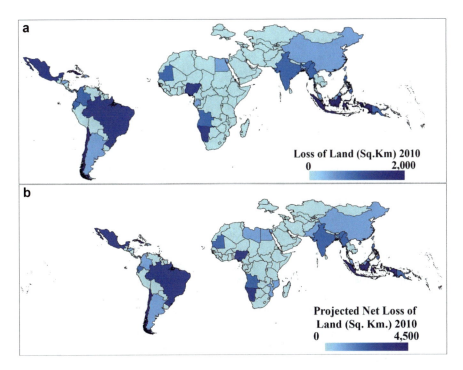

Fig. 5.2 Loss of land to sea level rise in (**a**) 2010; (**b**) estimated for 2030 (Data Source: DARA 2017)

dent on these water bodies for their livelihood. It is estimated that the costs of sea level rise for some of the island nations located in the Pacific will cost 20% of their total GDP in 2030. Though there are limited comprehensive data on long term sea levels for all the individual small islands in the tropics, but there are documented case studies revealing the adverse impacts of climate change on some of these islands. However, it is important to note that some of these islands have also undergone changes in the human settlement patterns in the form of increased urbanization and industrialization that have resulted in higher levels of erosion and subsidence. For instance, aggregate mining was found to be a major cause for accelerated beach erosion in the case of Anjouan Island, Comores in the Indian Ocean (Sinane et al. 2010). Similarly, the effects of anthropogenic activities in the form of expansion of human settlements in the Majuro atoll of the Marshall Islands obscure the direct effects of sea level rise (Ford 2012). Inadequate management practices resulting in beach erosion have also been cited in many of the other small islands in the tropics (Restrepo 2012).

Overall, the impacts of sea level rise are the greatest on SIDS, which in some cases threaten their survival as a nation. Therefore, the populations of these threatened islands are in some cases considered as the "early adaptors" (Betzold 2015). In most cases they are forced to migrate to other islands or seek asylum in larger neighboring countries. Due to the lack of data on climate change induced migration of population from small islands, the empirical relationship between the effects of anthropogenic climate change and migration is not widely validated. Specific case studies in the Pacific Islands indicate the role of environmental change on land use and land rights that have triggered migration (Bedford and Bedford 2010). In this context, the Pacific Access Category of migration that has been formulated between New Zealand and Tuvalu, allows 75 Tuvaluans to migrate to New Zealand every year. This agreement is supposed to promote economic and social migration as part of the Pacific Island's lifestyle (Shen and Gemenne 2011; Kravchenko 2008). The situation may be more extreme in the case of Kiribati, where the President of Kribati is in talks with the government of Fiji to buy up to 5000 acres of land to resettle the population from Kiribati (Chapman 2012). The migration or displacement of population is not only related to long-term increases in sea levels, but also the impacts of extreme weather events such as hurricanes in the Caribbean, and an unusually high inundation event in 2008 in Papua New Guinea (McLeman and Hunter 2010; Jarvis 2010). In some cases, the spread of extensive saltwater intrusion has rendered the soil unfit for agriculture leading to the decline in the cultivation of the staple food crops, and thus leading to forced displacement and malnutrition among the local population. For instance, it is projected that a 1 m rise in sea levels will lead to a decrease of agricultural land by 7% in Vietnam (Dasgupta 2009) and significant decrease in the rice yield, the staple crop of Myanmar, Bangladesh, and the Mekong River Delta (Wassmann et al. 2009). At the regional level, Bangladesh is considered to be the most vulnerable to the impacts of sea level rise, due to the estimated 60% of the population at risk due to contamination to fisheries, farmland, and salinization of drinking water sources. Here about half of the population lives less than 5 m above sea level on the world's fastest growing delta formed by the Ganges and Brahmaputra rivers. It is estimated that a 2 m rise in sea level will displace more than seven million residents in the Mekong Delta region, which is projected to double in case of a 2 m increase. In 1995, half of Bhola Island in Bangladesh was permanently flooded resulting in 500,000 people becoming homeless. In some of the projected estimates for 2030, about 20 million people will be displaced from their homes in Bangladesh due to flooding of their land related to sea level rise (Wax 2007).

Additionally, there is substantial evidence of climate sensitive impacts on the health of the local populations of some of the islands, in the form of morbidity and mortality, resulting from vector and water-borne diseases due to extreme weather events. One of the common manifestations include the contamination of drinking water supply after major tropical cyclones leading to the spread of water-borne diseases such as cholera, as observed in Eastern India after cyclone Alia in 2009 and Southern India after cyclone Thane in 2012. Most of the tropical islands are also considered to be generally favorable for the spread of vector borne diseases like malaria and dengue, and leptospirosis, which is projected to increase as a result of the impacts of sea level rise.

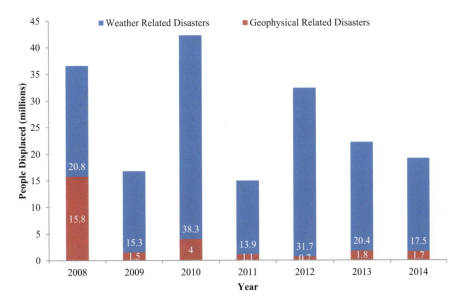

Fig. 5.3 Disaster-induced displacement. Source: IDMC estimates as of 01 June 2015 (rounded figures)

Extreme Weather Events

The increasing levels of complexity and frequency of displacement of population due to extreme weather events or weather-related disasters has been documented widely across the world. Disasters are defined by the serious disruption of the normal functioning of a community resulting in widespread human, material, economic, or environmental damages. These events usually exceed the local capacity of the population to cope using existing resources (UNISDR 2016). These displacements of population are usually temporary, but can become permanent in some cases, when the original places are made uninhabitable as a result of the disaster. Majority of the population displaced by extreme weather events often occur within national borders. Therefore, there is not much systematic reporting of population displacement at the international level. However, as evident from Fig. 5.3, the number of people displaced by extreme weather events constitutes a major proportion of IDPs. In 2014 alone, 17.5 million people were displaced due to disasters worldwide, which included weather and climate-related disasters such as floods, typhoons/cyclones, and storms. This was significantly lower compared to the two previous years, which were 20.4 million in 2013 and 31.7 million in 2012.

One of the common impacts of extreme weather disasters is in form of floods and flashfloods. The effects of floods are felt over a longer period of time due to the slower pace of rise and subsidence in water levels, while flashfloods occur within a very short time period, catching people unaware and unprotected. The impacts of both kinds of floods are more detrimental on women and girls (Nelson et al. 2002;

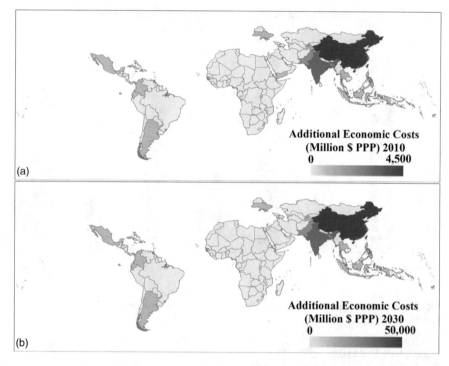

Fig. 5.4 Additional costs due to flooding (**a**) 2010; (**b**) 2030

Neumayer and Plümper 2007). In many of the countries in the Global South, due to the presence of clear gender roles, women and girls are often restricted to their homes. Majority of the women and girls do not know how to swim in case of floods, and often due to restrictive clothing have difficulty escaping inundated areas. For instance, during the 2010 Pakistani floods, more women experienced difficulty in accessing the relief distribution centers or were overlooked for relief distribution (Velasquez et al. 2016). Similarly, in Lagos and Nigeria, women in low income neighborhoods reported higher impacts and slower recovery from natural disasters such as floods (Ajibade et al. 2013).

Due to the higher density of population living in low-elevation, flood-prone areas on the coast, additional economic costs due to flooding will increase in the future (Fig. 5.4). The displacement of population will be the highest in Asia, particularly India, China, and Philippines, which accounted for 87% of population displaced by weather-related disasters. In addition, due to the limited collection of data on population displaced by natural disasters, there is limited documentation.

Food Insecurity

One of the direct impacts of the projected changes in climate conditions in the future is its impact on food security. Changing climate conditions are also expected to impact the quality and quantity of food production, which would lead to overall food shortages and rising food prices (Porter et al. 2014). The impacts of climate change and variability are again already more evident in the densely-populated vulnerable regions of the Global South, including South and Southeast Asia and Sub-Saharan Africa. The impacts of climate change will be greater on wheat and maize production in the tropics, while the higher latitudes, such as Northeast China or the UK, may benefit from a longer growing season (Jaggard et al. 2007; Chen et al. 2010). The impacts of climate change are not only limited to food produced on land but also on aquatic food supply from both freshwater and marine sources.

In the majority of the countries in the Global South, most of their economies are intricately tied with the performance of the agricultural sector. The majority of their populations live in the rural areas engaged in agriculture and livestock for their main sources of income and sustenance. The agricultural sector is also one of the most sensitive sectors to long-term climate change and short-term climate variability. Furthermore, it is hard to establish the extent of the impacts, because of the uncertainty associated with the local conditions related to resilience and adaptation. However, despite the uncertainty associated with the impacts of climate change, there is high confidence that major staple crops in the tropics, including rice, maize, and wheat, will be adversely impacted by projected climate change without any adaptation (IPCC 2014). In this context, in the lesser-developed areas, where mostly rain-fed agriculture is dominant, the impacts can be profound, compared to areas where there are other sources of irrigation such as canal or well irrigation that can reduce the severity of the impacts. Untimely rainfall or heat stress caused by extreme high temperatures or untimely frost can lead to extensive failures in crop yields and harvest. The results of crop yield modeling studies indicate that a local warming of a 1–2 °C in the tropics will result in the decline in wheat and maize yields (Porter et al. 2014). Furthermore, it is estimated that there will be geographic shifts and expansion of pests and diseases affecting crops due to rising temperatures and changes in precipitation regime (IPCC 2014). In addition, livestock grazing can also be affected by inadequate availability of fodder, which may force pastoralists to look for greener pastures in nearby agricultural lands. This puts the pastoralists in direct confrontation with local subsistence farmers.

Box 5.2 The Great Green Wall Initiative

It is a groundbreaking initiative to plant 4000 mile long wall of trees along the border of the Sahara Desert in the Sahel region to combat desertification, drought, and degradation. This program is based on the indigenous techniques of planting native trees and using simple indigenous water harvesting techniques. At 8000 km stretching across the width of the African continent, once completed it will be the largest living structure on the earth surface. This initiative is currently being implemented by 20 countries in Africa, with 8 billion dollars mobilized and/or promised in its support.

This wall has already boosted food security and resilience to climate change at the local scale. It may also help in the solution of some of the major problems faced in the region including famines, conflicts, and migration by empowering local population to thrive in their local environments (UNCCD 2017).

However, there has been substantial criticism of the plan due to the alteration of the natural desert ecosystems with variable evidence of it rapid expansion in Africa. Critics have also raised doubts about the pace and efforts for making the wall a viable reality. In order to implement the wall at across the proposed area, there is a need for the implementation of other sustainable techniques in addition to the farmer managed natural regeneration. Some of the suggested techniques include digging half-moons (to capture water) and planting seedlings (Lestadius 2017).

Additionally, in the rural areas with modernization and diversion of agricultural land to more lucrative cash intensive crops, small land holders are disproportionately disadvantaged. Many of these small land holders are female headed households, who are disproportionately affected over time. The impacts of food insecurity will be greatest in the lower latitudes, with some of the regional levels projected declines in yields at 23% for South Asia, 17% for East Asia and the Pacific, and 17% for Sub-Saharan Africa, and 14% for South America (Havlík et al. 2015). It is also noteworthy that the yield will not only decrease for the staple food crops, but also major cash producing crops, such as land areas available for cocoa production in Ghana and Côte d'Ivoire, and coffee production in Uganda (Läderach et al. 2013; Jassogne et al. 2013). Moreover, over the last ten years, 5 hurricanes have resulted in an average loss of about 10% of cultivable land in coastal Mexico, which have mainly affected single crop small farmers (WFP 2016a, b). The impact of climate change on agriculture related food insecurity would also impact the marine fisheries, with the potential redistribution of fishing around the world, in some cases overfishing. Furthermore, the impact of changes in sea surface temperatures, extreme weather events, and sea level rise will also affect marine ecosystems. The immediate impacts on fisheries will be most felt by the low lying coastal communities of small islands in the Pacific, as well as the major fishing nations of Chile, Peru, Colombia, and the tropical countries of Asia and Africa.

In this context, as mentioned in previous chapters, climate change will lead to greater reinforcement of already existing gender inequities and inequalities. In the poorer and more vulnerable countries, women are already in a disadvantaged position because of archaic laws in existence in certain countries, which limit or exclude them from land ownership rights. In many of these countries, larger numbers of women do subsistence agriculture. The spatial patterns of percentage of female

Food Insecurity

Table 5.1 Women and Men engaged in firewood collection and average time burden in a sample of countries in the Global South

Region	Year	Percentage collecting firewood Women	Men	Average time burden in population (minutes per day) Women	Men
Africa					
Benin	1998	22	5	16	4
Ghana	1998/1999	35	16	37	30
Madagascar	2001	10	15	7	13
Malawi	2004/2005			19	3
Morocco	1997/1998	3		3	
South Africa	2000	5	2	5	3
Asia					
Lao People's Democratic Republic	2002/2003			18	6
Pakistan	2007	4	2	3	2
Central America					
Nicaragua	1998	9	34	8	39

Sources: Compiled by the United Nations Statistics Division from World Bank, Gender, Time Use, and Poverty in Sub-Saharan Africa (2006) and time use survey reports from national statistical offices of Lao People's Democratic Republic, Nicaragua, Pakistan and South Africa

Note: Average time burden in population is calculated taking into account those involved in firewood collection as well as those not involved. Data may not be strictly comparable across countries as the methods involved for data collection may differ

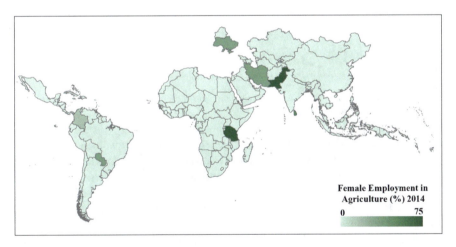

Fig. 5.5 Distribution of female employment in agriculture sector for percentage of female employment in 2014

employment in the agriculture sector, consisting of hunting, forestry and fishing, during 2014 are shown in Fig. 5.5. The countries with the highest percentage of female employment in the agricultural sector were Pakistan (74%) and Tanzania (70%), followed by Gambia (38.3%), Sri Lanka (33.9%), Iran (21.8%), and

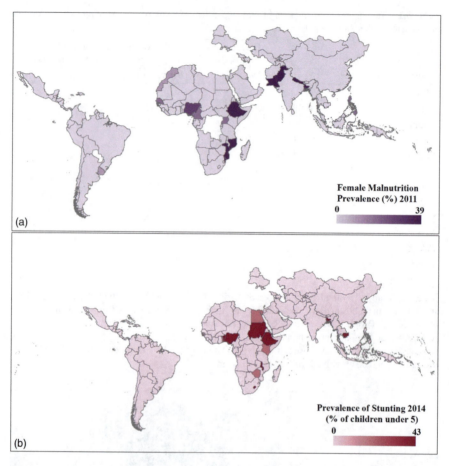

Fig. 5.6 Distribution of (**a**) Prevalence of female malnutrition in 2011; (**b**) Prevalence of stunting of children under 5 years of age

Paraguay (16.6%). However, in many of these societies even if women are the main caretakers of the land, they often do not have ownership rights. In addition, because of limited access to resources, bank credit, and decision making organizations, they often do not have knowledge or access to better quality seeds and advanced techniques of farming. For instance, in-depth research in Africa indicates that less than 10% of the credits granted to small farmers are given to women (Nair et al. 2006). The limited access to credit puts them at a disadvantage to buy new varieties of seeds or any modern equipment or extensions for agriculture. Thus, they will be most vulnerable to the impacts of climate change. The number of female-headed households will increase in poverty in rural areas, often caused by severe droughts that result in the out migration of men from the rural areas to the urban in search of alternate employment.

According to the UN in certain parts of Africa, women and girls spend up to 8 h collecting water for their daily needs, which is projected to increase with intensifying water scarcity in the future (Women's Earth & Climate Action Network International 2016). In addition, a substantial part of the day is spent on collecting firewood, which is again much greater for women and girls compared to men (Table 5.1). For example, in some of the countries in Sub-Saharan Africa countries, including Ghana and Benin, and Central America, including Nicaragua, women spend double to quadruple times the amount of time collecting firewood. Due to the lack of gender disaggregated data available on the amount of time spent collecting water and firewood, it is difficult to estimate the actual impacts of climate change on human well-being, which will only worsen as a result of rising temperatures and deficient precipitation.

Based on model projections, it is estimated that by 2050 climate change will lead to a 3.2% per person reduction in global food availability, which includes 4.0% reduction in fruits and vegetables consumption (Springmann et al. 2016). This reduced food supply may also lead to deaths as a result of undernourishment, dietary changes, with the maximum impacts experienced in the densely populated regions of Western Pacific and Southeast Asia, particularly China and India. It is estimated that by 2080, an additional 35 million to 170 million people in developing countries will be malnourished. At the regional scale, estimated 25–90% increases in the rates of malnourishments are projected by 2050 due to a 2 °C warming in Sub-Saharan Africa (Environmental Justice Foundation 2016). The latest figures for female malnutrition at the national level are mapped in Fig. 5.6. More than 20% of the female population in several countries with semi-arid climate located in the Sub-Saharan Africa had malnutrition, which included Nigeria, Cameroon, and Ethiopia. Similar higher levels of malnutrition among females were also reported in South Asia, including Pakistan (30.4%), Nepal (28.4%), and Bangladesh (39%). Some of the common characteristics among these countries are their high vulnerability to impacts of climate change in the form of prolonged and more frequent droughts, including Pakistan, Ethiopia, and Mozambique, and extreme weather, such as hurricanes, and sea level rise in Bangladesh. In addition, all of these countries already have large gender inequalities, which may lead to an increase in the levels of malnutrition due to the shortage of basic food supplies. Furthermore, the physiological needs of pregnant and lactating women also make them and children less than five years more susceptible to malnutrition. Maternal health is crucial for the health of the newborn baby and their survival. It is estimated that globally 50% of all pregnant women are anemic, the majority of whom live in the less developed regions of the Global South (Ransom and Elder 2016). Also, mostly women have limited access to medical facilities in the rural areas due to sparse network. Such as in Northwestern India, which is predominantly arid, women in the rural areas experience high levels of malnutrition, and as a result of limited access to medical facilities they also have high pregnancy and miscarriage rates. As shown in Fig. 5.6, the prevalence of childhood stunting due to malnutrition is more widespread than female malnutrition, with some of them overlapping for both categories, such as Ethiopia and Nigeria in Africa, and Bangladesh in South Asia. Additionally, food

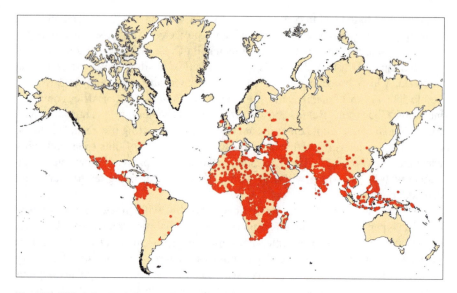

Fig. 5.7 Global distribution of conflicts during 1989–2010

insecurity will lead to higher food prices, which will make it less affordable for the poorer sections of the society to fulfil their nutritional needs. It is important to resolve women's malnutrition, because of the multiple positive impacts that is would benefit her health and also help in lowering child mortality, the overall health of the family. This would result ultimately in greater gender equality through better access to resources. In view of the greater proportion of women living under poverty in the Global South, therefore the effects of food shortage will be felt disproportionately by women and girls in the Global South.

Conflicts

One of the long term impacts of the effects discussed above will be the forced displacement and resulting in out migration of affected population from their homes to nearby areas within and across national borders. In many cases the out migration or displacement of population will be the result of conflicts. There is substantial evidence linking the indirect and direct impacts of climate change causing conflicts in various parts of the Global South. A report by the Christian Aid estimated that approximately 1 billion migrants will be displaced from their original homes due to issues related to food shortages and water (Christian Aid 2007). Additionally, the latest report of the IPCC specifically highlights the impacts of climate change on increased risks of violent conflicts in the form of civil wars and inter-group violence through negative impacts on poverty and economic shocks. The impacts of climate change will also be evident on critical infrastructure and territorial integrity (IPCC

2014). Despite large uncertainties associated with attributing climate change to conflicts, there are regional level studies showing the direct and indirect impacts of climate change on conflicts. The propensity of conflicts occurring due to climate change related processes are greater in economically less-developed and politically unstable regions of the world, which typifies the Global South. Therefore, even though direct causality is difficult to establish in some cases, several recent studies have indicated the role of climate change driven food insecurity on the occurrence or increased frequency of conflicts in certain regions of the world. The occurrences of conflicts in the mid-1950s have increased steadily, with most of them concentrated in the Global South (Fig. 5.7). However, not all of these conflicts can be attributed to food insecurity or natural disasters caused by climate change.

However, there is substantial literature examining the role of climate change on conflicts. Several studies have shown the impact of climate change related scarcity of resources on the occurrence of conflicts and local violence in the Global South. In this context, Darfur was widely considered to be the first conflict caused by climate change and variability, which led to mass killings in the African continent (Mazo 2009). However, later studies have indicated that local and regional level political and social factors played a much more prominent role in the Darfur conflict (Kevan and Gray 2008; Hagan and Kaiser 2011; Verhoeven 2011; Mazo 2009). However, the propensity of climate change related processes resulting in resource scarcity that can lead to armed conflicts in various regions of the Global South have been emphasized. For instance in Nigeria the role of climate change in forcing people of different ethnic and religious beliefs to live together have resulted in armed conflicts and violence at the local scale (Folami and Folami 2013). More generalized analysis of conflicts due to climate change across the African continent showed the positive relation between warmer temperatures and conflicts between 1980 and 2012 (Loughlin et al. 2014). The threats of climate change on the increased incidences of violence and common crime resulting from displacement related to food insecurity have been reported in Central and South America. Specifically, Honduras, El Salvador, and Guatemala are considered to be the poorest and most food insecure countries in Central and South America (FAO 2014). Along with food insecurity there is also large economic inequality and suppression of basic human rights, which makes the effects of food insecurity and hunger on the low income population more acute. There is chronic under-nutrition close to 50% in Guatemala as a result of food insecurity related to single crop farming, famines, and poor access to land for women (WFP 2016a, b). This has led to a mass exodus of population from these countries northwards toward the US. There are widespread reports of the increased levels of violence in the origin areas, but also on the migrator routes as they try to escape to better conditions. It is also noteworthy that the impact of climate change has not only been in the recent past, but studies document historical evidence of climate patterns such as ENSO on civil conflicts around the world (Davis 2002; Fagan 2009). Analysis of more recent data from 1950 to 2004 revealed the role of ENSO on 21% of all civil conflicts (Hsiang et al. 2011). In addition, future model predictions of higher temperatures indicate an approximately 54% increase in armed conflicts (Burke et al. 2009). It is estimated that approximately 46

110 5 Climate Refugees

countries, which consists of 2.7 billion population, are considered to be at risk of violent conflict due to the combined effects of climate change and ongoing socio-economic and political problems. Most of these countries are located in the Global South. Additionally, it is estimated, by 2050 there will be an increase in interpersonal violence by about 8–16%, and intergroup conflict by 28–56% (EJF 2016). The reported and projected increase in conflicts and interpersonal violence again puts women and girls in a vulnerable position, as evidenced by the increasing trends in gender-based violence during armed conflicts. There is increasing evidence of gender-based violence being committed by family members and people from their own communities.

Box 5.3 Syrian Crisis

In a recent set of published studies in major peer reviewed journals and popular media highlight the role of climate change on the Syrian conflict. Before the Syrian conflict began in 2011, this region experienced the most severe drought since instrumental record keeping started. Syria being a poorly governed nation with unsustainable agricultural and environmental policies, made it conducive for the breakdown in order and political unrest. The collapse of agriculture resulted in large scale migration of population from the agricultural rural areas to the larger cities, which is estimated to be about 1.5 million people (Integrated Regional Information Networks 2009; Solh 2010). Analysis of climate data reveals that the recent decrease in precipitation in Syria is the result of natural variability and long term drying trends (Kelley et al. 2015). This drought started in the winter of 2006–2007, which aggravated the already existing water and agricultural insecurity in the region. This region has a long history of conflicts and political unrests for most of its history. Many of these conflicts have been over water due to the relative shortage of water in this region, making it a precious commodity in majority of the power conflicts in this region. Particularly, in case of Syria, it has been argued that factors related to severe droughts, persistent multiyear crop failures, water shortage and mismanagement led to the deterioration of social structures in place (Femia and Werrell 2012; Mhanna 2016). In addition it also led to urban unemployment and economic dislocations, which finally led to widespread social unrest (Gleick 2014). Analysis of long term climate data revealed the role of anthropogenic forces on the increasing severity and persistence of droughts in this region (Kelley et al. 2015). Some of the mismanagement measures by the Syrian Government include the subsidizing of water intensive crops such as wheat and cotton and encouragement of inefficient irrigation techniques. The National Agricultural Policy Center, a research institution of the Syrian Ministry of the Agriculture and Agrarian Reform, reported an increase in underground wells into the ground aquifers from just over 135,000 in 1999, to more than 213,000 in 2007. This resulted in the lowering of the groundwater table in major parts of the country, with concerns about the

References 111

quality of the limited water left in the aquifers (Femia and Werrell 2012; Mhanna 2016). The conditions leading up to the Syrian conflict and its impacts on the wider society highlights the indirect and direct impacts of climate change not only in the form of conflicts, but also food insecurity and forced migration from rural to urban areas and finally across borders. The Syrian conflict has resulted in a major international humanitarian crisis rendering people homeless many miles away from their home country, stuck in refugee camps and often dangerous living conditions. Latest estimates of Syrian population forced to leave their country is 4 million in five host countries, with more than 16 million people in need of assistance inside and outside Syria. The five major host countries as of February 5, 2016 include Turkey (1.93 million), Lebanon (1.17 million), Jordan (0.63 million), Iraq (0.25 million), and Egypt (0.13 million). More than 11 million people have been forced to leave their homes or killed (Mercy Corps 2016). The Syrian crisis highlights the effects of forced migration aggravated by climate change leading food insecurity and ultimately major violent conflicts.

The effects of this crisis are not only limited to the five host countries and Syria, but has generated widespread global discussions about humanitarian aid and treatment of refugees. It is widely considered to be the worst humanitarian crisis since World War II. It is estimated by the UN that more than half of the Syrian refugees are children under the age of 18 years and women. Thus they are losing out on their education, which makes their future very uncertain, referred to as the "lost generation". According to the UN High Commissioner for Refugees (UNHCR), approximately 100,000 children have been born in refugee camps with no birth certificates. In many cases the youngest children are growing up confused and scared with lack of any form of social order or security, while the older children are forced to fend for themselves and their families by finding odd jobs. A deeper analysis of interviews and anecdotal reports from the various refugee camps and inside war torn Syria, reveals the higher vulnerability of women and girls to gender based violence, including rape, kidnappings, and forcible marriages. There is not much data available on gender based violence, because in most instances victims of domestic violence, mostly women and children do not seek for help outside the family.[1]

---

[1] UN Women Inter-Agency Assessment, Gender-Based Violence and Child Protection Among Syrian Refugees in Jordan, With a Focus on Early Marriage, Jordan, July 2013, p. 28, and Child Protection and Gender Based Violence Sub-Working Group Jordan, Findings from the Inter-Agency Child Protection and Gender-Based Violence Assessment in the Za'atari Refugee Camp, Jordan, February 2013, p. 3.

References

Ajibade, I., McBean, G., & Bezner-Kerr, R. (2013). Urban flooding in Lagos, Nigeria: Patterns of vulnerability and resilience among women. *Global Environmental Change, 23*, 1714–1725.

Becker, M., Meyssignac, B., Letetrel, C., Llovel, W., Cazenave, A., & Delcroix, T. (2012). Sea level variations at tropical Pacific islands since 1950. *Global and Planetary Change, 80-81*, 85–98.

Bedford, R., & Bedford, C. (2010). International migration and climate change: A post-Copenhagen perspective on options for Kiribati and Tuvalu. In B. Burson (Ed.), *Climate change and migration: South Pacific perspectives*. Wellington: Institute of Policy Studies.

Betzold, C. (2015). Adapting to climate change in small island developing states. *Climatic Change, 133*, 481–489.

Brown, O. (2016). Climate change and forced migration: Observations, projections and implication. *A background paper for the 2007 Human Development*. Retrieved March 6, 2016, from https://www.iisd.org/pdf/2008/climate_forced_migration.pdf

Burke, M., Miguel, E., Satyanath, S., Dykema, J. A., & Lobell, D. B. (2009). Warming increases the risk of civil war in Africa. *Proceedings of the National Academy of Sciences, 106*, 20670–20674.

Chapman, P. (2012). Entire nation of Kiribati to be relocated over rising sea level threat. *The Telegraph*. Retrieved March 7, 2012, from http://www.telegraph.co.uk/news/worldnews/australiaandthepacific/kiribati/9127576/Entire-nation-of-Kiribati-to-be-relocated-over-rising-sea-level-threat.html

Chen, C. E., Wang, E., Yu, Q., & Zhang, Y. (2010). Quantifying the effects of climate trends in the past 43 years (1961-2003) on crop growth and water demand in the North China Plain. *Climatic Change, 100*, 559–578.

Christian Aid. (2007). *Human tide: The real migration crisis* (A Christian Aid Rep.). London.

Conisbee, M., & Simms, A. (2003). *Environmental refugees: The case for recognition*. New Economics Foundation, London. Retrieved March 6, 2016, from www.neweconomics.org/publications/environmental-refugees

DARA. (2017). *Climate vulnerability monitor*. Retrieved November 15, 2017, from http://daraint.org/climate-vulnerability-monitor/climate-vulnerability-monitor-2012/data/indicator-data/?area=climate&indicator=Sea-Level%20Rise

Dasgupta, S. (2009). The impact of sea level rise on developing countries: A comparative analysis. *Climatic Change, 93*, 379–388.

Davis, M. (2002). *Late Victorian Holocausts: El Niño famines and the making of the third world*. London: Verso.

EJF. (2014). *The gathering storm: Climate change, security and conflict*. London: Author.

EJF. (2016). *Climate and conflict*. Retrieved June 7, 2016, from http://ejfoundation.org/campaigns/oceans/item/climate-and-conflict

Environmental Justice Foundation. (2016). *Protecting people and planet: Climate and conflict*. Retrieved May 26, 2016, from http://ejfoundation.org/campaigns/oceans/item/climate-and-conflict

Erkens, G. (2016). Subsidence in megacities on the coast greater than absolute sea level rise. *Deltares*. Retrieved March 26, 2016, from https://www.deltares.nl/en/news/subsidence-megacities-coast-greater-absolute-sea-level-rise/

Fagan, B. (2009). *Floods, famines and emperors: El Niño and the fate of civilizations*. New York: Basic Books.

FAO. (2014). *Panorama de la Seguridad Alimentaria y Nutricional en Centroamerica y Republica Dominicana 2014*, Panama: Author.

Femia, F., & Werrell, C. (2012, February 29). Syria: Climate change, drought, and social unrest. *The Center for Climate and Security*. Retrieved February 17, 2016, from http://climateandsecurity.org/2012/02/29/syria-climate-change-drought-and-social-unrest/

References

Folami, O. M., & Folami, A. O. (2013). Climate change and inter-ethnic conflict in Nigeria. *Peace Review, 25*, 104–110.

Ford, M. (2012). Shoreline changes on an urban atoll in the central Pacific Ocean: Majuro Atoll, Marshall Islands. *Journal of Coastal Research, 28*, 11–22.

Gleick, P. H. (2014). Water, drought, climate change, and conflict in Syria. *Weather, Climate, and Society, 6*, 331–340. https://doi.org/10.1175/WCAS-D-13-00059.1.

Hagan, J., & Kaiser, J. (2011). The displaced and dispossessed of Darfur: Explaining the sources of a continuing state-led genocide. *British Journal of Sociology, 62*, 1–25.

Hallegate, S., Green, C., Nicholls, R. J., & Corfee-Morlot, J. (2013). Future flood losses in major coastal cities. *Nature Climate Change, 3*. https://doi.org/10.1038/nclimate1979.

Havlík, P., Valin, H. J. P., Gusti, M., Schmid, E., Forsell, N., Herrero, M., et al. (2015). *Climate change impacts and mitigation in the developing world: An integrated assessment of the agriculture and forestry sectors* (World Bank Policy Research Working Paper No. 7477).

Hijioka, Y. E., Lin, J. J. P., Corlett, R. T., Cui, X., Insarov, G. E., Lasco, R. D., et al. (2014). Asia. In V. R. Barros, C. B. Field, D. J. Dokken, M. D. Mastrandrea, K. J. Mach, T. E. Bilir, et al. (Eds.), *Climate change 2014: Impacts, adaptation, and vulnerability. Part B: Regional aspects. Contribution of working group II to the Fifth Assessment Report of the Intergovernmental Panel on Climate Change*. Cambridge: Cambridge University Press.

Hsiang, S. M., Meng, K. C., & Cane, M. A. (2011). Civil conflicts are associated with the global climate. *Nature, 476*, 438–441.

Hsiang, S. M., Burke, M., & Miguel, E. (2013). Quantifying the influence of climate on human conflict. *Science, 341*. https://doi.org/10.1126/science.1235367.

Integrated Regional Information Networks. (2009, November 24). Syria: Drought response faces funding shortfall. *IRIN*. Retrieved February 17, 2016, from http://www.irinnews.org/report/87165/syriadrought-response-faces-funding-shortfall

IPCC. (2014). Summary for policymakers. In C. B. Field, V. R. Barros, D. J. Dokken, K. J. Mach, M. D. Mastrandrea, T. E. Bilir, et al. (Eds.), *Climate change 2014: Impacts, adaptation, and vulnerability. Part A: Global and sectoral aspects. Contribution of working group II to the Fifth Assessment Report of the Intergovernmental Panel on Climate Change*. Cambridge: Cambridge University Press.

Jaggard, K., Qi, A., & Semenov, M. A. (2007). The impact of climate change on sugarbeet yield in the UK: 1976-2004. *The Journal of Agricultural Science, 145*, 367–375.

Jarvis, R. (2010). Sinking nations and climate change adaptation strategies. *Seattle Journal of Social Justice, 9*, 447–486.

Jassogne, L., Läderach, P., & Van Asten, P. (2013). The impact of climate change on coffee in Uganda: Lessons from a case study in the Rwenzori Mountains. *Oxfam Policy and Practice: Climate Change and Resilience, 9*, 51–66.

Kelley, C. P., Mohtadi, S., Cane, M. A., Seager, R., & Kushnir, Y. (2015). Climate change in the Fertile Crescent and implications of the recent Syrian drought. *Proceedings of the National Academy of Sciences, 112*, 3241–3246.

Kevan, M., & Gray, L. (2008). Darfur: Rainfall and conflict. *Environmental Research Letters, 3*. https://doi.org/10.1088/1748-9326/3/3/034006.

Kravchenko, S. (2008). Right to carbon or right to life: Human rights approaches to climate change. *Vermont Journal of Environmental Law, 9*, 513–547.

Laczko, F., & Aghazarm, C. (Eds.). (2009). *Migration, environment and climate change: Assessing the evidence*. Geneva: International Organization for Migration.

Läderach, P., Martinez-Valle, A., Schroth, G., & Castro, N. (2013). Predicting the future climatic suitability for cocoa farming of the world's leading producer countries, Ghana and Côte d'Ivoire. *Climatic Change, 119*, 841–854.

Lestadius, L. (2017). *Africa's got plans for a Great Green Wall: Why the idea needs a rethink*. Retrieved October 30, 2017, from http://theconversation.com/africas-got-plans-for-a-great-green-wall-why-the-idea-needs-a-rethink-78627

Loughlin, J. O., Linke, A. M., & Witmer, D. W. (2014). Effects of temperature and precipitation variability on the risk of violence in sub-Saharan Africa, 1980–2012. *Proceedings of the National Academy of Sciences, 111*, 16712–16717.

Mazo, J. (2009). Chapter five: Climate change and security. *The Adelphi Papers, 49*, 119–136.

McLeman, R. A., & Hunter, L. M. (2010). Migration in the context of vulnerability and adaptation to climate change: Insights from analogues. *Wiley Interdisciplinary Reviews: Climate Change, 1*, 450–461.

Mercy Corps. (2016). Quick facts: What you need to know about the Syria crisis. *Mercy Corps*. Retrieved February 5, 2016, from https://www.mercycorps.org/articles/iraq-jordan-lebanon-syria-turkey/quick-facts-what-you-need-know-about-syria-crisis

Mhanna, W. (2016). *Syria's climate crisis*. Retrieved February 17, 2016, from http://www.al-monitor.com/pulse/politics/2013/12/syrian-drought-and-politics.html#

Myers, N. (2016). Environmental refugees: An emergent security issue. In *13th economic forum, May 2005, Prague*. Retrieved March 6, 2016, from www.osce.org/documents/eea/2005/05/14488-en.pdf

Nair, S., Sexton, S., & Kirbat, P. (2006). A decade after cairo women's health in a free market economy. *Indian Journal of Gender Studies, 13*, 171–193.

Nelson, V., Meadows, K., Cannon, T., Morton, J., & Martin, A. (2002). Uncertain predictions, invisible impacts, and the need to mainstream gender in climate change adaptations. *Gender and Development, 10*, 51–59.

Neumayer, E., & Plümper, T. (2007). The gendered nature of natural disasters: The impact of catastrophic events on the gender gap in life expectancy, 1981–2002. *Annals of the Association of American Geographers, 97*, 551–566.

Porter, J. R., Xie, L., Challinor, A. J., Cochrane, K., Howden, S. M., Iqbal, M. M., et al. (2014). Food security and food production systems. In *Climate change 2014: Impacts, adaptation and vulnerability, Contribution of working group II to the Fifth Assessment Report of the Intergovernmental Panel on Climate Change*. Cambridge: Cambridge University Press.

Ransom, E. I., & Elder, L. K. (2016). *Nutrition of women and adolescent girls: Why it matters*, Population Reference Bureau. Retrieved June 3, 2016, from http://www.prb.org/Publications/Articles/2003/NutritionofWomenandAdolescentGirlsWhyItMatters.aspx

Restrepo, J. C. (2012). Shoreline changes between 1954 and 2007 in the marine protected area of the Rosario Island Archipelago (Caribbean of Colombia). *Ocean & Coastal Management, 69*, 133–142.

Rhein, M., Rintoul, S. R., Aoki, S., Campos, E., Chambers, D., Feely, R. A., et al. (2013). Observations: Ocean. In T. F. Stocker, D. Qin, G.-K. Plattner, M. Tignor, S. K. Allen, J. Boschung, et al. (Eds.), *Climate change 2013: The physical science basis. Contribution of working group I to the Fifth Assessment Report of the Intergovernmental Panel on Climate Change*. New York: Cambridge University Press.

Shen, S., & Gemenne, F. (2011). Contrasted views on environmental change and migration: The case of Tuvaluan migration to New Zealand. *International Migration, 49*, e224–e242.

Sinane, K., David, G., Pennober, G., & Troadec, R. (2010). Fragilisation et modification des formations littorales meubles sur l'île d'Anjouan (Comores): Quand l'érosion d'origine anthropique se conjugue au changement climatique. *Vertigo, La Revue Électronique en Sciences de l'Environnement, 10*.

Solh, M. (2010, September 27). Tackling the drought in Syria. *Nature Middle East*. Retrieved February 17, 2016, from https://www.natureasia.com/en/nmiddleeast/article/10.1038/nmiddleeast.2010.206

Springmann, M., Mason-D'Croz, D., Robinson, S., Garnett, T., Godfray, H. C. J., Gollin, D., et al. (2016). Global and regional health effects of future food production under climate change: A modelling study. *The Lancet, 387*, 1937–1946.

Stern, N. (2007). *The economics of climate change: The Stern review*. Cambridge: Cambridge University Press.

Syvitski, J. P. M. (2009). Sinking deltas due to human activities. *Nature Geoscience, 2*, 681–686.

5 Climate Refugees

Teatini, P., Castelletto, N., Ferronato, M., Gambolati, G., Janna, C., Cairo, E., et al. (2011). Geomechanical response to seasonal gas storage in depleted reservoirs: A case study in the Po River basin, Italy. *Journal of Geophysical Research: Earth Surface, 116*, F02002. https://doi.org/10.1029/2010JF001793.

The Nansen Initiative. (2016). *Disaster induced cross border initiative*. Retrieved March 16, 2016, from https://www.nanseninitiative.org/

UNCCD. (2017). *The Great Green Wall initiative*. Retrieved October 30, 2017, from http://www2.unccd.int/actions/great-green-wall-initiative

UNGA. (2016). *United Nations General Assembly (UNGA) Resolution A/63/L.8/Rev.1 - Climate change and its possible security implications. 2009*. Retrieved March 5, 2016, from http://www.un.org/esa/dsd/resources/res_pdfs/ga-64/cc-inputs/Iceland_CCIS.pdf

UNISDR. (2016). *Terminology on DRR*. Retrieved March 29, 2016, from http://www.unisdr.org/we/inform/terminology

Velasquez, J., Sirimanne, S., Bonapace, T., Srivastava, S. K., Mohanty, S., et al. (2016). *Reducing vulnerability and exposure to disasters: The Asia-Pacific Disaster Report 2012, ESCAP UNISDR*. Retrieved April 4, 2016, from http://www.unisdr.org/files/29288_apdr2012finallowres.pdf

Verhoeven, H. (2011). Climate change, conflict and development in Sudan: Global Neo-Malthusian narratives and local power struggles. *Development and Change, 42*, 679–707.

Wassmann, R., Jagadish, S. V. K., Sumfleth, K., Pathak, H., Howell, G., Ismail, A., et al. (2009). Regional vulnerability of climate change impacts on Asian rice production and scope for adaptation. *Advances in Agronomy, 102*, 91–133.

Wax, E. (2007). In flood-prone Bangladesh, a future that floats. *The Washington Post*. Retrieved September 27, 2007, from http://www.washingtonpost.com/wp-dyn/content/article/2007/09/26/AR2007092602582.html

WECAN. (2016). *Why women are key*. Retrieved March 6, 2016, from http://wecaninternational.org/pages/why-women-are-key

WFP. (2016a). *Hunger without borders*. Retrieved June 6, 2016, from http://reliefweb.int/sites/reliefweb.int/files/resources/wfp277544.pdf

WFP. (2016b). *Food insecurity and climate change*. Retrieved June 02, 2016, from www.wfp.org/climate-change

Women's Earth & Climate Action Network International. (2016). *Why women are key*. Retrieved May 26, 2016, from http://wecaninternational.org/pages/why-women-are-key

Wong, P. P., Losada, I. J., Gattuso, J.-P., Hinkel, J., Khattabi, A., McInnes, K. L., et al. (2014). Coastal systems and low-lying areas. In C. B. Field, V. R. Barros, D. J. Dokken, K. J. Mach, M. D. Mastrandrea, T. E. Bilir, et al. (Eds.), *Climate change 2014: Impacts, adaptation, and vulnerability. Part A: Global and sectoral aspects. Contribution of working group II to the Fifth Assessment Report of the Intergovernmental Panel on Climate Change*. New York: Cambridge University Press.

Chapter 6
Resilient and Sustainable Cities

Sustainable Development Goal 6: "Make cities inclusive, safe, resilient and sustainable" (UN 2016).

Currently, it is estimated that about 54% of the total human population live in urban areas, which is a 30% increase from 1950. In view of the rapid rates of urbanization and predictions of 66% of the world population living in urban areas by 2050, the impact of climate change related processes on the human civilization have become more critical. The majority of the urban expansion and population increase are concentrated in the developing world, with nearly 90% of the increase predicted to occur in Asia and Africa. The percentage proportion of population living in urban areas is shown in Fig. 6.1a. Most of the population in South America and the Middle East lived in urban areas in 2015, while the lowest proportion of urban population were concentrated in Central and Eastern Africa, Asia, and the Pacific Island nations in Oceania. Specifically, 23 countries in the Global South had less than 25% of their population living in urban areas, which included 9 in Africa, 7 in Oceania, 4 in South America and the Caribbean, and 3 in Asia. It is noteworthy that six of the ten countries with the highest proportion of urban population (almost 100% urban population) are located in the Global South, which include Singapore, Qatar, Kuwait, and Hong Kong, Macau in China, Guadeloupe, and Uruguay are all located in the Global South. The spatial patterns of the trends in percentage of urban population from 1950 to 2015 indicate the largest increase in percentage urban population in the northern part of South America, North and East coast of Africa, Middle East, and East and Southeast Asia (Fig. 6.1b). All countries in the Global South experienced positive trends in the growth of urban population except Tajikistan in Central Asia and some of the small island countries located in the Caribbean, including Trinidad and Tobago, Belize, Barbados, Antigua and Barbuda, and Guyana. It is projected that by 2050 more than half of the population will be in urban areas in 80% of the countries worldwide, with the maximum increase observed in Asia (56%) and Africa (64%). Most of the increase in urban population will be concentrated in the Global South.

© Springer International Publishing AG, part of Springer Nature 2018
S. Sen Roy, *Linking Gender to Climate Change Impacts in the Global South*,
Springer Climate, https://doi.org/10.1007/978-3-319-75777-3_6

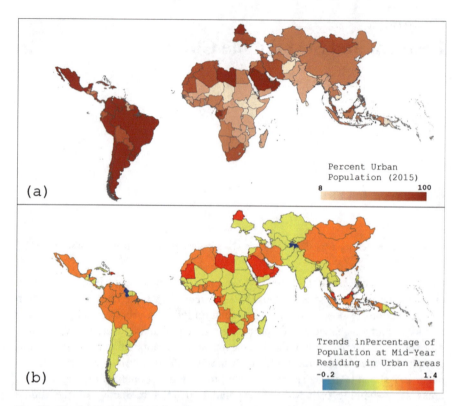

Fig. 6.1 Distribution of Urban Population at the country level (**a**) Percentage of Urban Population in 2015; (**b**) Trends in percentage of population at mid-year. (Data Source: UN 2014)

These increased rates of urbanization along with a rapid increase in urban population in Asia and Africa, are often a combination of both push and pull factors. In most of Asia and Africa, cities are often associated with better employment and educational opportunities which attracts lot of younger population from adjacent rural areas. Besides the pull factors there is overcrowding and declining yield in some cases in the agricultural areas, which have resulted in the forced migration of people to nearby cities in search of employment. This forced migration initially consists of the younger male population who can do manual jobs in the cities. In some cases this rural to urban migration is seasonal, consisting of younger men moving to the city for gainful employment for a few months of the year and then returning to their villages at harvest time to help out in the farms. However, due to over population in the rural areas, lot of the younger men stay on in the cities.

This increasing trend in rural-urban migration has led to the creation of an increasing size of transient population in some of the large metropolitan areas of Africa and Asia, including Lagos, Nairobi, Mexico City, Buenos Aires, New Delhi, Dhaka, and Manila. This is further magnified by the vast disparities between the tier one cities and cities of a lower order in the Global South. Figure 6.2 summarizes the

6 Resilient and Sustainable Cities

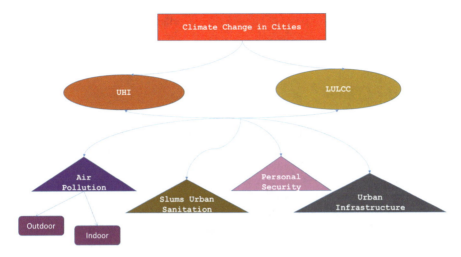

Fig. 6.2 Brief Summary of the direct and indirect impacts of climate change related processes in cities

direct and indirect impacts of climate change related processes in urban areas of the Global South. Unlike the spatial patterns of urbanization in the developed world, where all cities have certain level of amenities and infrastructure, most of the tier one cities, often the administrative capital or financial hub of a country in the Global South have disproportionately larger infrastructure than the cities in the next tier. This happens mainly due to unbalanced distribution of resources and infrastructure development from the national governments in most of these countries. As a result, there is not only rural to urban migration, but also migration from smaller cities to the larger cities. This results in even more overcrowding in the big cities, and exerts pressure on the already thinly stretched resources and infrastructure in these cities. As a result, most of the migrants end up living in slums with the bare amenities and unhygienic conditions. Therefore, the growths of slums or favelas or shanty towns or townships have increasingly become an integral part of many of the large metropolitan areas in the Global South.

The latest estimate of people living in slums is 828 million, which is rising rapidly (UN 2016). These areas are characterized by high unemployment and crime rates, along with substandard living conditions. In addition, there is also a wide cultural disconnect between the rural areas and the large cities in the Global South. One of the impacts of this wide cultural disconnect along with the presence of large numbers of transient population in the cities has resulted in lower levels of personal security, particularly for women and girls. This is particularly evident from the increased harassment of female population in large cities of the Global South, with the worst manifestation being in the form of the rape incident that took place in December 2012 in New Delhi. The rest of the chapter is an in-depth discussion about the various issues identified so far in this chapter in the context of already occurring and potential impacts of climate change related processes as well as gendered differences.

Urban Heat Island

In the context of urban climate, one of the distinct characteristics of urban areas include the occurrence of relatively higher temperatures and in some cases extreme weather events in large urban areas compared to adjacent rural areas. These differences in weather conditions between urban and rural areas are defined as the Urban Heat Island (UHI) effect. This phenomenon is considered to be one of the most distinct and widely validated anthropogenic impacts of climate change on the local environment. UHI development is mainly caused by the concentration of heat absorbing surfaces, such as concrete buildings and asphalt streets. The high density of tall buildings in the core urban areas disturbs the air flow patterns, leading to the creation of the urban canyon effect and the modification of the surrounding environment. Additionally, the effect of the radiative trapping of heat due to building geometry, anthropogenic heat sources such as automobile combustion, energy consumption in the high rise buildings, and greater urban surface roughness, results in reduced surface albedo and greater rates of heat absorption (Arnfield 2003). Thus the UHI effect is typically characterized by lower diurnal temperature ranges and higher rate of increase in surface temperatures.

However, in the latest report of the IPCC, the effect of UHI phenomena was declared to be more localized, with negligible impact on large scale climate trends (Hartmann et al. 2013). Yet, in view of more than half of the world's population living in urban areas, the effects of UHI are far reaching. Additionally, the modulating impact of urbanization on local level weather phenomena on the downwind side of several large urban centers includes higher temperatures, reduced air quality, and modification of the hydrological cycle (Bohnenstengel 2011). This is further intensified by the rapidly increasing population in some of the large metropolitan areas, where it has resulted in not only horizontal expansion into adjacent farmlands, but also vertical expansion in the form of high rise residential buildings. This densification of urban areas results in the trapping of heat and pollutants in the urban canopy layer, which have detrimental impacts on the overall quality of life. The higher temperatures in urban areas have led to greater levels of consumption of energy and other basic amenities.

The presence of well-developed UHI effect is more widely validated in temperate mid-latitude cities (Bärring et al. 1995; Brazel et al. 2005; Giannaros et al. 2013; Hathway and Sharples 2012). However, recent studies in large metropolitan areas located in the tropics and subtropics indicate the distinct presence of the UHI effect, such as that observed in Singapore, New Delhi, Bangkok, and Beijing (Akbari et al. 2016; Sen Roy et al. 2011; Arifwidodo and Tanaka 2015; Song et al. 2014). This is particularly significant in the Global South where most of the increase in urban population and urbanization has been concentrated over the last few decades

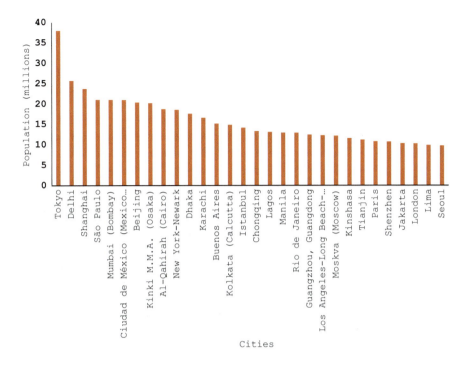

Fig. 6.3 Population in 30 largest urban agglomerations in 2015. (Data Source: UN 2014)

(Fig. 6.3). Among the 30 largest urban agglomerations on the global scale, 24 are located in the Global South. More specifically, only three cities in the top ten largest cities (Tokyo, Osaka, and New York) were not located in the Global South, which is predicted go down to only one city (Tokyo) in 2030. Among the 23 largest cities in the Global South in 2015, 12 are capital cities, and about half the cities (11) are located inland, while the rest of the cities (13) are located on the coast. Many of these densely populated cities located in the coastal areas are already experiencing the negative impacts of rising sea levels, such as Manila, Mumbai, and Lagos. The cities that are located inland, on the other hand, are more prone to the impacts of UHI effect, such as evidenced in New Delhi, Beijing, and Mexico City. For instance a review of literature attributed the impact of UHI and land use land cover change (LULCC) related processes at 20% of overall trends in Eastern China, and about 0.1 °C per decade at the national level in China (Hartmann et al. 2013). Given current trends in population demographics, these trends will probably continue for the next few decades at the same rate. In addition it is important to mention that the rate of urbanization in the Global South has occurred at a much faster pace than the developed world, which has caused excessive stress on existing amenities and infrastructure in the Global South.

Often the urban expansion in many of these cities has taken place without much planning into adjacent agricultural areas. Furthermore, there is a big gap between the largest one or two cities at the national level and smaller cities in the Global South. In most of the countries in the Global South the rapid pace of development are usually concentrated in the top two or three largest urban agglomerations. In many of these countries the most significant presence of UHI are located in the capital cities of the more densely populated countries, including China, India, Mexico, Argentina, and Democratic Republic of Congo.

There has been an increase in specific case studies about some of these large urban agglomerations in the Global South in have increased since 2000, which have highlighted the development and characteristics of UHI in some of these cities. The expansion of research about UHI analysis in the Global South has been facilitated by the availability of satellite images over an extended period of time and adequate spatial resolution. The majority of the studies focusing on UHI analysis are usually the administrative or financial capitals of their countries, including Beijing and Shanghai in China, New Delhi and Mumbai in India, Sao Paulo and Rio de Janeiro in Brazil, and Mexico City in Mexico.

Some of the dominant themes involving the analysis of UHI development in the large urban agglomerations of the Global South include elevated levels of air pollution, occurrence of extreme temperatures in the urban core compared to adjacent rural areas, absence of green vegetation, and the rapid urban sprawl in adjacent rural areas. Additionally, the relation between population size and the development of UHI is widely established for mid-latitude cities (Oke 1973, 1981), but there is a general lack of any comprehensive studies in the Global South examining the role of growing population directly on UHI. This is critical particularly in the Global South, which has been experiencing a burgeoning population at a rapid pace. For example, Mexico City, like many other capital cities in the Global South has an inland location surrounded my mountain which enhances the UHI effect along with the impact of anthropogenic activities resulting from an increase in population from 3.37 million in 1950 to 21 million in 2015. The projected population for Mexico City in 2030 is 23.86 million. The effect of an increasing population further intensifies the negative impacts of anthropogenic activities which results in the intensification of UHI.

There are substantial spatial and temporal variations associated with UHI development at the local scale. For example, a detailed analysis of hourly data collected from a network of weather stations in Beijing indicated the manifestation of a pronounced UHI during night time hours in the winter months (Zhong et al. 2015). At the spatial scale, the intensity of the UHI was highest in the city core. Similar seasonal level differences in the intensity of UHI development, particularly during the nocturnal hours of winter months, have also been reported in other large urban centers in the Global South including Shanghai, New Delhi, Seoul, and Sao Paulo (Jiang et al. 2004; Ferreira et al. 2012). However, the reverse stronger development of UHI in summer season was found in Buenos Aires, Dar es Salaam, and Ouagadougou (Camilloni and Barrucand 2012; Jonsson 2004; Jonsson et al. 2002). Additionally, surface level UHI was more intense during daytime in Southern China, while the reverse was observed in Northern China (Wang et al. 2015). Also, the role of UHI on the intensification of extreme weather events, mainly convective thunderstorms and floods has been found in many large urban centers. For instance, a 200% increase in lightning activity was observed in Bogota, Colombia compared to adjacent rural areas. The role of urbanization on the increased occurrence of extreme precipitation events was also found in Beijing and major urban areas over the Indian subcontinent (Song et al. 2014; Sen Roy 2009). The role of urbanization showed an overall decline in summer and dry month precipitation in Beijing, while an increase in precipitation was observed in Kolkata (Song et al. 2014; Mitra et al. 2012). Thus the impact of urbanization even at the local level is evidently complex and can have far reaching impacts on the majority of the world's population if unchecked.

Land Use and Land Cover Changes (LULCC)

One of the distinct physical manifestations of urbanization on the local landscape is the change in land use and land cover, which alters the local level energy balance. With the advent of satellite imagery that has higher resolution than before, it has become possible to monitor LULCC with much more detail. In this regard, several studies have examined the impact of urbanization on the reduction of vegetation in urban areas. This is particularly significant for large urban agglomerations in the Global South, where unplanned urban sprawl is typically associated with the conversion of adjacent agricultural land to urban land uses. In many of the large cities of the Global South, many open areas demarcated for parks or green areas are covered by illegal squatter settlements. In addition, the conversion of green spaces to urban built-up land uses result in greater heat retention and long term negative consequences, unless active efforts are made to combat the negative impacts by using reflective roof or rooftop gardens among other measures. The local level LULCC have a direct impact on surface and near surface UHI development. For instance, in the case of Mexico City, the land surface and near surface UHI was found to be high at night during all seasons, except the wet season. However, there were differences in the day time surface and near surface UHI levels. Particularly, the UHI development was found to be more strongly correlated with the vegetation fraction than the daily insolation (Cui and De Foy 2012).

More specifically, the type of land use and land cover have a direct impact on the prevailing land surface temperatures (LST), which in turn impacts the UHI intensity. For example, the changes in LST over the last sixteen years in three major urban centers (Beijing, Cairo, and Mexico City) of the Global South have been mapped using MODIS satellite imageries (Fig. 6.4). All of these cities experienced the spread of higher temperatures from the core to the outskirts of the city, which may mostly be attributed to conversion from rural to suburban areas to urban land uses such as asphalt and concrete buildings, often referred to as impervious surfaces. In fact, one study shows there is evidence of a direct correlation between impervious surface area and LST observed in many of the cities in China (Zhang et al. 2009; Zhou et al. 2010). In addition, the expansion in impervious surface area

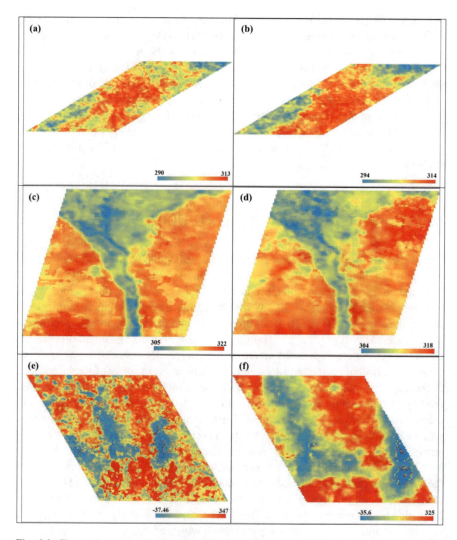

Fig. 6.4 Change in summer daytime Land Surface Temperature (LST) in Kelvin between 2000 and 2016 extracted from MODIS satellite images (**a**) Beijing 2000; (**b**) Beijing 2016; (**c**) Cairo 2000; (**d**) Cairo 2016; (**e**) Mexico 2000; (**f**) Mexico 2016

also results in increased runoff and evapotranspiration rates, leading to inadequate recharge of groundwater tables.

Another major impact of unplanned urbanization in some of the large urban agglomerations in the Global South is the reduction in green areas, which aggravates the negative impacts of UHI effect in the form of higher levels of sensible heat flux. The amount of vegetation in urban areas is commonly measured by the widely used greenness index, Normalized Vegetation Index (NDVI) derived from satellite imageries. The relationships between NDVI and surface temperatures have been widely examined in a variety of locations. In Beijing there was a strong correlation between LST and NDVI, while a weak correlation between the two variables were found in Fuzhou City in China (Xiao et al. 2008; Zhang et al. 2009). Overall, vegetated areas are usually associated with lower LSTs in urban areas (Grover and Singh 2015). However, there is general consensus about the decline in vegetation cover as a result of urbanization and its negative impact on the overall air quality and ambient temperatures. Specifically, the positive role of large scale afforestation on cooling of LST was found in China, through increased carbon storage, and alteration of the local albedo and turbulent energy fluxes (Peng et al. 2014).

Some of the direct impacts of LULCC are on the localized occurrence of extreme temperatures resulting in severe heat waves, flashfloods as a result of decreased infiltration of rainwater, poorer air quality, and extended impacts in the downwind areas of some of the large urban centers.

Air Pollution

An inevitable impact of the unplanned rapid urban sprawl in and around major urban centers in the Global south is the deterioration of the physical environment. Due to the substantially higher per capita consumption practices in urban areas, it is estimated that approximately two-thirds of the world's energy consumption and 70% of the global greenhouse gas emissions occur in cities (UN-Habitat 2011). Specifically, increased income levels among urban residents have led to greater levels of per capita ownership of motor vehicles, which has resulted in widespread congestion, decrease in open and public spaces, and decreased urban air quality. In many cases, the lower income populations are disadvantaged due to the lack of adequate public transportation and forced to live closer to the sources of pollution. Typically, in many of the large cities of the Global South there are increasing trends in particulate matter, ozone, and other greenhouse gases such as carbon monoxide and nitrogen oxide. Latest estimates from the World Health Organization (WHO) indicate that more than 80% of residents in urban areas are at risk of exposure to air pollution. According to the urban air quality data base, 98% of cities in low and middle income countries which have more than 100,000 residents exceed the WHO air quality guidelines (WHO 2016). This creates critical health concerns for urban residents in the low and middle income countries.

One of the earliest analyses of UHI development for major cities in the Global South was conducted for Mexico City, which highlighted the increase in local levels of turbidity and global radiation. Since this analysis, there have been additional studies highlighting the elevated levels of urban air pollution as a result of the impact of UHI development, which traps the pollutants inside the urban canopy layer. Additionally, the impact of urban sprawl and aerosol levels on long term trends in overall and extreme precipitation patterns were found for some of the urban stations (Jauregui 1958; Oke et al. 1999; Baumgardner and Raga 2016). Elevated levels of air pollution in the urban boundary layer have also been reported

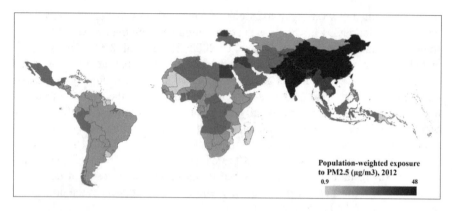

Fig. 6.5 Spatial patterns of weighted population exposure to PM 2.5 in 2012

in several other large urban agglomerations in the Global South including Beijing, New Delhi, Hong Kong, and Buenos Aires.

Air pollution in cities is mostly caused by different anthropogenic processes including traffic, industry, power plants, and domestic fuels. Among these sources, traffic has been considered as the main source of urban pollutants, as evident from the high levels of pollutants near busy traffic intersections and highways, which include nitrogen oxides and ozone (Liu and Sen Roy 2014; O'Shea et al. 2015). Pollutants emitted in urban areas are dispersed and diluted, which in turn is determined by the prevailing meteorological conditions, local surface morphology, and topographical settings. For instance, due to lesser turbulence in the atmosphere, the levels of pollutants are usually higher during the winter months and night time due to calmer conditions. In fact, two of the most polluted cities in world to be in the news for high levels of air pollution are Beijing and New Delhi, which also are the capital cities of the two most populous nations in the world.

Relatively lower air quality in urban areas is a result of the interaction of physical environment and anthropogenic activities, which has become a critical issue in most of the cities in the Global South. In the latest report on air quality levels for particulate matter (PM 10 and PM 2.5), the top 100 cities in descending order of air quality are located in the Global South. Onitsha in Nigeria was the most polluted city, followed by Peshawar in Pakistan, Zabol in Iran, Rawalpindi in Pakistan, Kaduna and Aba in Nigeria, Riyadh and Al Jubail in Saudi Arabia, Mazar-e-Sharif in Afghanistan, and Gwalior, in India representing the top ten most polluted cities. It is not surprising that most of these cities are located in the arid and semi-arid climates, which makes them more prone to elevated levels of dust and fine particulate matter. At the regional scale, all the regions in the Global South with adequate available data showed an increasing trend in the levels of PM 2.5 and PM 10 between 2008 and 2013.

The health impacts of the elevated levels of particulate matter in the environment are particularly detrimental for human health through the deep penetration of these particles into the respiratory tract. It increases the chances of respiratory diseases such as asthma, lung cancer, and increased mortality from respiratory infections and

Air Pollution

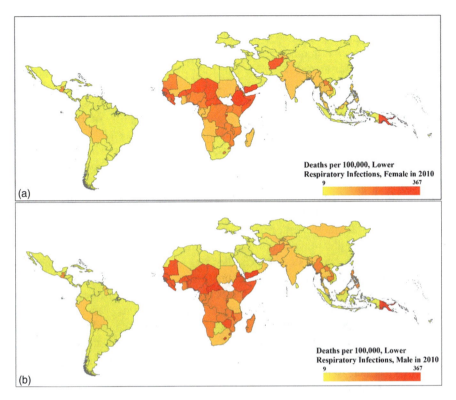

Fig. 6.6 Spatial patterns of deaths resulting from lower respiratory infectious diseases in 2010 per 100,000 population (**a**) female; (**b**) male

cardiovascular diseases. According to WHO reports in 2012, about 7 million people (1 in 8 of total global deaths) were related to exposure to ambient (3.7 million deaths) and household (4.3 million deaths) air pollution (WHO 2014). The spatial patterns of the number of people affected and mortality caused by air pollution in 2010 across the Global South is shown in Fig. 6.5. Not surprisingly, countries with higher density of population are more affected by air pollution, including China, India, and most of Southeast Asia. A recent study reveals air pollution from coal sources as the largest source of air pollution-related health impacts in China. Specifically, exposure to ambient fine particulate matter (PM 2.5) led to 916,000 premature deaths in 2013 (Health Effects Institute 2016). These countries are also characterized by a rapid pace of economic development and industrialization, which has resulted in excessively low air quality in many of the large cities of Asia. The patterns are also consistent with the long term trends in regional air pollution levels observed by WHO, which indicates increasing trends in air pollution levels across all the regions in the Global South. In addition to pollutants, the impact of climate change will also be felt in the quantity, quality, timing, and duration of pollens in urban areas.

The role of air pollution on human health is widely established. From the overall spatial patterns of deaths resulting from lower respiratory infectious diseases are represented in Fig. 6.6. It is clearly evident that a greater number of female deaths per 100,000 population occurred in parts of Asia, Afghanistan and Oman, (Fig. 6.6a), though the overall the number of deaths caused by air pollution were higher and more widespread for men (Fig. 6.6b). This may also be partially due to the greater number of males per 100,000 total populations. However, the burden of air pollution on women is greater than on men. One of the main reasons for women and girls to be particularly vulnerable is because they constitute the majority of the poorer and disadvantaged sections of the population exposed to higher levels of air pollution. More specifically, women outnumber the elderly populations who have weaker aging immune systems exposed to the long term effects of air pollution. Recent scientific evidence shows the lifetime exposure to estrogens a key cause for breast cancer. More specifically, the presence of certain chemical agents in the environment can affect the production and metabolism of estrogen within the body (Melius et al. 1994). Additionally, hormonal changes during pregnancy, lactation, and menopause can result in the internal storage of pollutants in the body, which can affect health many years later. For instance, dioxins, a group of chemical compounds released in the environment by industrial processes such as manufacturing, combustion, and chlorine bleaching of pulp and paper, play a role in the development of disease and illness among women (Bryant 1996). Another study revealed the impact of exposure to air pollutants, such as nitrogen oxides, carbon monoxide, sulfur dioxide, and ozone, on pregnant women in the form of gestational diabetes mellitus leading to adverse impacts on birth outcomes (Robledo et al. 2015). Also, studies assessing the increased risk of lung cancer in urban areas, revealed a greater risk of cardiovascular diseases among postmenopausal women in urban areas due to long term exposure to particulate matter (Miller et al. 2007).

Air pollution is not only related to outdoor pollution, but also indoor air pollution, where many women spend their days doing household chores. It is estimated that about 3 billion people worldwide depend on solid fuels, including biomass (wood, dung, and agricultural residues), and coal for their most basic energy needs. The spatial patterns of Disability Adjusted Life Years (DALY) per 100,000 of population attributed to household air pollution from solid fuels show a substantial clustering over South Asia and Sub Saharan Africa (Fig. 6.7). The inefficient burning of these solid fuels often results in the formation of a hazardous combination of pollutants, consisting of carbon monoxide and small particles nitrogen oxides, benzene, butadiene, formaldehyde, polyaromatic hydrocarbons and many other health-damaging chemicals. Specifically, the burning of solid fuels produces extremely high levels of indoor air pollution, such as the typical 24-h levels of PM 10 in homes using biomass in Africa, Asia or Latin America range from 300 to 3000 $\mu g/m^3$. The peak levels during cooking can sometimes reach 10,000 $\mu g/m^3$.

Therefore, it is estimated that women and children, who spend hours in such levels of very poor indoor air quality, inhale amounts of smoke approximately equal to two packs of cigarettes per day (WHO 2017). This is mostly case for low income families, who cannot afford cleaner sources of energy. For instance, household indoor air pollution is one of the leading risk factors for air pollution in South and Southeast Asia for women and children, where the use of indoor stoves for cooking

Air Pollution

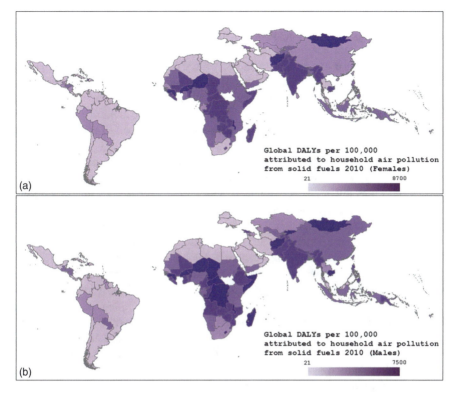

Fig. 6.7 Spatial patterns of DALYs per 100,000 population attributed to household air pollution from solid fuels (**a**) females; (**b**) males

family meals involve the use of solid fuels. It is the top risk factor for women in Cambodia (Mishamandani 2015). Inhaling smoke from burning solid fuels doubles the risk of pneumonia and other acute infections of the lower respiratory tract for children under 5 years, and triple the risk for women suffering from chronic obstructive pulmonary disease (COPD). Also the use of coal, which is a widely used source of fuel in homes in the Global South, doubles the risk of lung cancer among women. Indoor smoke related pollution also leads to an increase in infant mortality rates, with 16,400 fewer surviving infants in Indonesia (Jayachandran 2006).

Food Insecurity

The negative impacts of the rapidly occurring unplanned urbanization are not only limited to the urban centers, but in some cases they can also have widespread impacts on the peri-urban or adjacent rural areas. There are several case studies highlighting the contamination of heavy metals and other pollutants in the agricultural soil, which get transferred in vegetables and ultimately end up in the human system. The pathway of entry of heavy metals into the human system through this soil-crop system has long been considered as a major conduit. There is also widespread evidence of the anthropogenic source of heavy metals in the environment, which include: vehicular, industrial, and domestic emissions, weathering of buildings, pavements, and asphalt (Sezgin et al. 2003; Ahmed and Ishiga 2006). More specifically, the sources of heavy metals in agricultural areas originating from urban areas include urban effluent, vehicle emissions, and sewage (Yang et al. 2009; Montagne et al. 2007; Li et al. 2008).

In the recent years, there have been multiple case studies analyzing the contamination of heavy metals, originating in cities, in urban and adjacent agricultural soils for various cities in China. The review of literature for urban soils and urban road dusts show widespread contamination levels for lead, zinc, chromium, copper, and nickel. The contamination levels were expectedly highest in the larger tropical climate cities of Eastern China, including Hong Kong, Shanghai, Guangzhou, and Hangzhou. These contamination levels are just not limited to urban areas but also in agricultural soil near urban areas, particularly for lead, cadmium, and mercury. All of the heavy metals found in the agricultural soil in China are derived from anthropogenic activities (Wei and Yang 2010). For instance soil sample tested at different sites in Shunyi, an agricultural suburb of Beijing and suburbs of Nanjing in China, showed the presence of mercury caused by the atmospheric deposits from Beijing City (Lu et al. 2012, Wang et al. 2013). Urban contamination of heavy metals in agricultural systems has also been found in urban areas of other countries in the Global South. For instance, in the peri-urban areas of Sanandaj, Iran the levels of chromium and lead were much higher than safety limits specified by the Food and Agricultural Organization (FAO) (Maleki et al. 2014). Similar occurrences of heavy metals contamination, including lead, copper, and nickel, were found in vegetables grown in industrial and semi-urban areas of Central India and different regions of Pakistan (Pandey and Pandey 2014; Waseem et al. 2014).

Contaminations from anthropogenic activities in urban areas are not just limited to the peri-urban or semi-urban areas, but it also impacts vegetables in urban gardens extensively. In many cases, due to limited open areas agricultural activities are often concentrated alongside major streets, abandoned land that maybe exposed to toxic chemicals, and in some cases crowded informal housing such as slums or squatters. For instance, in Sao Paulo, Brazil, a strong correlation was found in the levels of trace metal composition in leafy vegetables and vehicular traffic burden (Amato-Lourenco et al. 2016). Similarly, higher levels of heavy metals concentration, including lead, cooper, zinc, and cadmium, in leafy vegetables grown near the international airport of Cotonou, Benin in Western Africa and in urban areas of northern Nigeria (Uzu et al. 2014; Egwu and Agbenin 2013).

It is noteworthy, that despite the negative impacts related with urbanization on soil and produce, there are substantial positive impacts of urban gardening. For instance urban gardening can help in reducing the UHI intensity at the micro level, in addition to providing fresh locally grown produce for the residents. Thus, it can promote better health among citizens along with social integration and environmental sustainability.

Additionally, urban gardening in many of the developing countries have helped in social inclusion and reduction of gender inequalities because in the majority of the urban farmers are women (Orsini et al. 2013). Some of the estimates from different urban areas in the Global South include more than 21,000 hectares of land in Cagayan de Oro City, Philippines, about 12% of the land in Havana, Cuba, and more than 11,000 hectares of land in Jakarta, Indonesia dedicated to agriculture (Potutan et al. 2000; Cruz and Medina 2003; Purnomohadi 2000). Most of the urban agriculture includes orchards, fruits, and vegetables. Furthermore, urban agriculture is becoming more widespread in the Global South with increasing proportions of urban population engaging in agriculture related activities in their backyards or potted plants on rooftops and balconies.

A substantial proportion of the urban population in many of the African cities are engaged in some type of agricultural activity which include 50% in Accra, Ghana, 80% in Brazzaville, Congo, 68% in the five biggest cities of Tanzania, 45% in Lusaka, Zambia and 37% in Maputo, Mozambique (Orsini et al. 2013). Urban agriculture allows the residents to grow organic vegetables and protect themselves from market fluctuations. Furthermore, several studies have highlighted the positive effects of urban agriculture on the lives of women and children. In a lot of traditional patriarchal societies, women are often the managers of these urban gardens, thus providing them an opportunity to gain entrepreneurship skills and expand their social support network through various agricultural community garden initiatives. It is estimated that about 65% of urban farmers worldwide are women (Van Veenhuizen 2006). In addition urban gardens can provide unique opportunities for children and youth to spend time outside their homes in a safe environment and learn traditional skills from their elders. Thus, there are significant benefits or urban gardening if properly managed.

However, as mentioned above women are the main drivers of urban agricultural activities and are also the major section of the labor force on the rural farms. While the effects of contamination of agricultural produce may not necessarily be gender sensitive, that is it affects both men and women who consume the contaminated agricultural produce. However, in view of majority of the urban agriculture activities conducted by women, puts them at a greater risk to the exposure to contaminants in the environment. For instance, in the Global South open areas are often not the most desirable areas for agriculture because often these areas are used as dumping grounds for trash and other solid wastes due to limited sanitation and trash collection services in large urban areas. These areas are disproportionately concentrated in the lower income areas near slums and squatters, who usually do not have much choice in terms of locating themselves. In some cases, the agricultural areas may also be areas which are former junkyards, meaning the presence of industrial by-products that are harmful to human health. Additionally, there are limited monitoring of urban agricultural produce and sites for contamination in the Global South. With the exception of limited studies conducted in differently sized urban centers across the Global South, there are no broad based initiatives from the local governments to monitor the levels of contamination in the local soil or the agricultural produce.

Personal Security in the Context of Changing Social Landscapes

One of the indirect impacts of climate change related processes is on cultural and social landscapes, which will vary across societies and time. This is mainly caused by the increased migration of mostly younger male population from the adjacent rural areas to the urban areas. This has resulted in compromised personal security in many of large urban areas. In many instances, it is very hard for the new immigrants to the city to find respectable means of livelihood, shelter, and in some cases sufficient food for survival. In most of the cases they end up living in squatter settlements and slums with very poor hygiene and infrastructure. Many of them end up compromising their cultural practices, identity, community cohesion, and sense of place. The majority of these slums or squatter settlements are located in the most undesirable parts of the city, which are prone to floods as well as irregular or unimproved sources of drinking water and sanitation. The latest data from the WHO indicates that only a third of the poorest urban households have piped water supply to their homes with the lowest in levels in Africa (GHO 2016).

Due to improper sewage treatment, often there is open sewage near homes which can be active breeding grounds for various water and air borne infectious diseases outbreaks. For instance, after Hurricane Matthew, as a result of the destruction of sanitation and water infrastructure there has been a surge in cholera outbreak in Port-au-Prince, Haiti. This is a widely prevalent scenario in many of the large metropolitan areas of the Global South. When there is an extreme rainfall event, the water often stagnates in the low lying areas of the cities where the slums are located. In many cases the water mixes with the open sewer or other water bodies and leads to water contamination and ideal breeding sites for mosquitos. The houses in these slums are often mostly made of very weak materials which are not able to endure heavy rainfall or standing flood waters. In addition, in these slums due to limited regulations from the urban authorities; there is a spur of antisocial behavior, such as gangs and other illegal activities.

According to the UN, 34 of the 39 countries with more than 50% of their urban population living in slums are located in Sub-Saharan Africa (Fig. 6.8) (Millennium Development Goals 2016). There has been substantial decrease in overall slum population, concentrated mostly in Asia. With the majority of projected increases in urban population occurring in Africa, the situation becomes even grimmer in the near future with already failing infrastructure. It is noteworthy that the majority of the increase in slum population has taken place in countries experiencing conflicts over the last two decades, such as Iraq with a 60% increase in slum population between 2000 and 2014 (UN 2016).

Furthermore, there has been a steep decline in personal security particularly for women and girls in the larger metropolitan areas. This decline in personal security can be attributed to the increased rural-urban migration, resulting in a substantial growth in the transient population in cities. Most of this transient population in the city lives in the slums without regular sources of employment. This is in contrast to

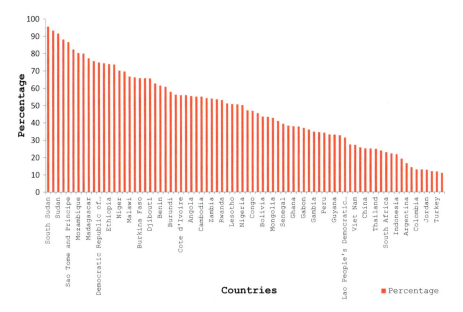

Fig. 6.8 Percentage of Urban Population Living in Slums in 2014

the very rich sections of the cities, where people have significant amount of money to spend. Over time the gap between high and low incomes has increased with a faster growth rate in the low income population. One of the indirect impacts is the increase in crime rates in large metropolitan areas of the Global South. Particularly, women and girls in large metropolitan areas are exposed to greater risks of sexual exploitation and abuse. In addition, there is a growing number of home invasions and attack on elderly people living by themselves. The majority of the instances, when the perpetrators are caught are usually recent migrants from nearby rural areas with no stable source of employment. In essence it is critical to make that as a result of climate change related processes large number of people are uprooted from their ancestral homes and forced to move to new surroundings. These new surroundings are significantly different from what they are accustomed to, resulting in major culture shock. Therefore, there is a critical need for the development of more explicit policies with greater coordination between stakeholders at different levels to prevent and respond to gender-based violence.

In this context, future projections indicate that trends in urbanization could produce a near tripling in the global urban land area between 2000 and 2030 (Angel et al. 2011; Seto et al. 2012). Therefore, as areas of land undergo changes in land cover and land use from non-urban to urban land uses, it will threaten or destroy habitats in key biodiversity hotspots. It will also contribute to increased anthropogenic carbon emissions associated with tropical deforestation and land use changes.

References

Ahmed, F., & Ishiga, H. (2006). Trace metal concentrations in street dusts of Dhaka city, Bangladesh. *Atmospheric Environment, 40*, 3835–3844.

Akbari, H., Cartalis, C., Kolokotsa, D., Muscio, A., Pisello, A. L., Rossi, F., et al. (2016). Local climate change and urban heat island mitigation techniques–the state of the art. *Journal of Civil Engineering and Management, 22*, 1–16.

Amato-Lourenco, L. F., Moreira, T. C. L., de Oliveira Souza, V. C., Barbosa, F., Jr., Saiki, M., Saldiva, P. H. N., et al. (2016). The influence of atmospheric particles on the elemental content of vegetables in urban gardens of Sao Paulo, Brazil. *Environmental Pollution, 216*, 125–134.

Angel, S., Parent, J., Civco, D. L., Blei, A., & Potere, D. (2011). The dimensions of global urban expansion: Estimates and projections for all countries, 2000–2050. *Progress in Planning, 75*(2), 53–107.

Arifwidodo, S. D., & Tanaka, T. (2015). The Characteristics of Urban Heat Island in Bangkok, Thailand. *Procedia-Social and Behavioral Sciences, 195*, 423–428.

Arnfield, A. J. (2003). Two decades of urban climate research: A review of turbulence, exchanges of energy and water, and the urban heat island. *International Journal of Climatology, 23*, 1–26.

Bärring, L., Mattsson, J. O., & Lindqvist, S. (1995). Canyon geometry, street temperatures, and urban heat island in Malmö, Sweden. *Journal of Climatology, 5*, 433–444.

Baumgardner, D., & Raga, G. (2016). *Changes in precipitation intensity in Mexico City: Urban Heat island effect or the impact of aerosol pollution?*. Retrieved July 15, 2016, from https://www.researchgate.net/profile/Graciela_Raga/publication/237197120_CHANGES_IN_PRECIPITATION_INTENSITY_IN_MEXICO_CITY_URBAN_HEAT_ISLAND_EFFECT_OR_THE_IMPACT_OF_AEROSOL_POLLUTION/links/0deec51ba4ea34e7b4000000.pdf

Bohnenstengel, S. (2011). Simulations of the London urban heat island. *Quarterly Journal of the Royal Meteorological Society, 137*, 1625–1640.

Brazel, A. J., Fernando, H. J. S., Hunt, J. C. R., Selover, N., Hedquist, B. C., & Pardyjak, E. (2005). Evening transition observations in Phoenix, Arizona. *Journal of Applied Meteorology, 44*, 99–112.

Bryant, K. (1996). Impact of air pollution on women's health. *Otolaryngology - Head and Neck Surgery, 114*, 267–270.

Camilloni, I., & Barrucand, M. (2012). Temporal variability of the Buenos Aires, Argentina, urban heat island. *Theoretical and Applied Climatology, 107*, 47–58.

Cruz, M. C., & Medina, R. S. (2003). *Agriculture in the city: A key to sustainability in Havana, Cuba*. Kingston: Ian Randle Publishers.

Cui, Y. Y., & De Foy, B. (2012). Seasonal variations of the urban heat island at the surface and the near-surface and reductions due to urban vegetation in Mexico City. *Journal of Applied Meteorology and Climatology, 51*, 855–868.

Egwu, G. N., & Agbenin, J. O. (2013). Field assessment of cadmium, lead and zinc contamination of soils and leaf vegetables under urban and peri-urban agriculture in northern Nigeria. *Archives of Agronomy and Soil Science, 59*, 875–887.

Ferreira, M. J., de Oliveira, A. P., Soares, J., Codato, G., Bárbaro, E. W., & Escobedo, J. F. (2012). Radiation balance at the surface in the city of São Paulo, Brazil: Diurnal and seasonal variations. *Theoretical and Applied Climatology, 107*, 229–246.

Giannaros, T. M., Melas, D., Daglis, I. A., Keramitsoglou, I., & Kourtidis, K. (2013). Numerical study of the urban heat island over Athens (Greece) with the WRF model. *Atmospheric Environment, 73*, 103–111.

Global Health Observatory (GHO) Data. (2016). *Access to piped water*. Retrieved December 19, 2016, from http://www.who.int/gho/urban_health/environmental-risk-factors/safe_water/en/

Grover, A., & Singh, R. B. (2015). Analysis of urban heat island (UHI) in relation to normalized difference vegetation index (NDVI): A comparative study of Delhi and Mumbai. *Environments, 2*, 125–128.

References

Hartmann, D. L., Klein Tank, A. M. G., Rusticucci, M., Alexander, L. V., Brönnimann, S., Charabi, Y., et al. (2013). Observations: Atmosphere and surface. In T. F. Stocker, D. Qin, G.-K. Plattner, M. Tignor, S. K. Allen, J. Boschung, et al. (Eds.), *Climate change 2013: The physical science basis. contribution of working group I to the Fifth Assessment Report of the Intergovernmental Panel on Climate Change 2013*. Cambridge: Cambridge University Press.

Hathway, E. A., & Sharples, S. (2012). The interaction of rivers and urban form in mitigating the Urban Heat Island effect: A UK case study. *Building and Environment, 58*, 14–22.

Health Effects Institute. (2016). *Burden of disease attributable to coal burning and other major sources of air pollution in China*. Retrieved August 2016, from https://www.healtheffects.org/system/files/GBDMAPS-ExecSummaryEnglishFinal.pdf

Jauregui, E. (1958). Increase in turbidity in Mexico City. *Ingenieria Hidraulica En Mexico, 12*, 1–10.

Jayachandran, S. (2006). Air quality and early life mortality: Evidence from Indonesia's wildfires. *Journal of Human Resources, 44*. https://doi.org/10.3386/w14011.

Jiang, T., Jiong, S., & Lian-tang, D. (2004). Wavelet characteristics of urban heat island in Shanghai City. *Journal of Tropical Meteorology, 5*, 006.

Jonsson, P. (2004). Vegetation as an urban climate control in the subtropical city of Gaborone, Botswana. *International Journal of Climatology, 24*, 1307–1322.

Jonsson, P., Eliasson, I., Lindqvist, S. (2002). Urban climate and air quality in tropical cities. In *Preprints fourth symposium on the urban environment*. Norfolk: AMS.

Li, Y., Gou, X., Wang, G., Zhang, Q., Su, Q., & Xiao, G. (2008). Heavy metal contamination and source in arid agricultural soils in central Gansu Province, China. *Journal of Environmental Sciences, 20*, 607–612.

Liu, Z., & Sen Roy, S. (2014). Spatial patterns of seasonal level diurnal variations of ozone and respirable suspended particulates in Hong Kong. *The Professional Geographer, 67*, 17–27.

Lu, A., Wang, J., Qin, X., Wang, K., Han, P., & Zhang, S. (2012). Multivariate and geostatistical analyses of the spatial distribution and origin of heavy metals in the agricultural soils in Shunyi, Beijing, China. *Science of the Total Environment, 425*, 66–74.

Maleki, A., Gharibi, F., Alimohammadi, M., Daraei, H., & Zandsalimi, Y. (2014). Concentration levels of heavy metals in irrigation water and vegetables grown in peri-urban areas of Sanandaj, Iran. *Journal of Advances in Environmental Health Research, 1*, 81–88.

Melius, J. M., Lewis-Michl, E. L., Kallenbach, L. R., Ju, C. L., Talbot, T. O., Orr, M. F., et al. (1994). *Residence near industries and high traffic areas and the risk of breast cancer on Long Island*. New York: New York Health Department.

Millennium Development Goals. (2016). *Indicators slum population as percentage of urban, percentage*. Retrieved December 19, 2016, from http://mdgs.un.org/unsd/mdg/seriesdetail.aspx?srid=710

Miller, K. A., Siscovick, D. S., Sheppard, L., Shepherd, K., Sullivan, J. H., Anderson, G. L., et al. (2007). Long-term exposure to air pollution and incidence of cardiovascular events in women. *New England Journal of Medicine, 356*, 447–458.

Mishamandani, S. (2015). Global burden of disease update reveals major risk factors for death and disability. *Science Spotlight*, from https://www.niehs.nih.gov/research/programs/geh/geh_newsletter/2015/12/spotlight/1215_spotlight_508.pdf

Mitra, C., Shepherd, J. M., & Jordan, T. (2012). On the relationship between the premonsoonal rainfall climatology and urban land cover dynamics in Kolkata city, India. *International Journal of Climatology, 32*, 1443–1454.

Montagne, D., Cornu, S., Bourennane, H., Baize, D., Ratié, C., & King, D. (2007). Effect of agricultural practices on trace-element distribution in soil. *Communications in Soil Science and Plant Analysis, 38*, 473–491.

O'Shea, P. M., Sen Roy, S., & Singh, R. B. (2015). Diurnal variations in the spatial patterns of air pollution across the Delhi Metropolitan Region. *Theoretical and Applied Climatology, 124*, 609–620.

Oke, T. R. (1973). City size and the urban heat island. *Atmospheric Environment, 7*, 769–779.

Oke, T. R. (1981). Canyon geometry and the nocturnal urban heat island: Comparison of scale model and field observations. *Journal of Climatology, 1*, 237–254.

Oke, T. R., Spronken-Smith, R. A., Jauregui, E., & Grimmond, C. S. B. (1999). The energy balance of central Mexico City during the dry season. *Atmospheric Environment, 33*, 3919–3930.

Orsini, F., Kahane, R., Nono-Womdim, R., & Gianquinto, G. (2013). Urban agriculture in the developing world: A review. *Agronomy for Sustainable Development, 33*(4), 695–720.

Pandey, R., & Pandey, S. K. (2014). Trace metal accumulation in vegetables grown in industrial and semi-urban areas of Singrauli District of Madhya Pradesh India. *International Journal of Pharmaceutical Science and Research, 12*, 5518–5529.

Peng, S.-S., Piao, S., Zeng, Z., Ciais, P., Zhou, L., Li, L. Z., et al. (2014). Afforestation in China cools local land surface temperature. *Proceedings of the National Academy of Sciences, 111*, 2915–2919.

Potutan, G. E., Schnitzler, W. H., Arnado, J. M., Janubas, L. G., & Holmer, R. J. (2000). Urban agriculture in Cagayan de Oro: A favourable response of city government and NGOs. In N. Bakker, M. Dubbeling, S. Gündel, U. Sabel-Koschella, & H. de Zeeuw (Eds.), *Growing cities growing food: Urban agriculture on the policy agenda* (pp. 413–428). Feldafing: DSE.

Purnomohadi, N. (2000). Jakarta: Urban agriculture as an alternative strategy to face the economic crisis. In N. Bakker, M. Dubbeling, S. Gündel, U. Sabel-Koshella, & H. de Zeeuw (Eds.), *Growing cities growing food: Urban agriculture on the policy agenda* (pp. 453–465). Feldafing: DSE.

Robledo, C. A., Mendola, P., Yeung, E., Männistö, T., Sundaram, R., Liu, D., et al. (2015). Preconception and early pregnancy air pollution exposures and risk of gestational diabetes mellitus. *Environmental Research, 137*, 316–322.

Sen Roy, S. (2009). A spatial analysis of extreme hourly precipitation patterns in India. *International Journal of Climatology, 29*, 345–355.

Sen Roy, S., Singh, R. B., & Kumar, M. (2011). An analysis of local-spatial temperatures patterns in the Delhi metropolitan area. *Physical Geography, 32*, 114–138.

Seto, K. C., Güneralp, B., & Hutyra, L. R. (2012). Global forecasts of urban expansion to 2030 and direct impacts on biodiversity and carbon pools. *Proceedings of the National Academy of Sciences, 109*(40), 16083–16088.

Sezgin, N., Ozcan, H. K., Demir, G., Nemlioglu, S., & Bayat, C. (2003). Determination of heavy metal concentrations in street dusts in Istanbul E-5 highway. *Environment International, 29*, 979–985.

Song, X., Zhang, J., AghaKouchak, A., Roy, S. S., Xuan, Y., Wang, G., et al. (2014). Rapid urbanization and changes in spatiotemporal characteristics of precipitation in Beijing metropolitan area. *Journal of Geophysical Research: Atmospheres, 119*, 11250–11271. https://doi.org/10.1002/2014JD022084.

UN. (2014). *World urbanization prospects: The 2014 revision*. CD-ROM Edition.

UN. (2016). *SDGs: Sustainable development knowledge platform. United Nations*. Retrieved December 29, 2016, from https://sustainabledevelopment.un.org/sdgs

UN-Habitat. (2011). *Hot cities: Battle-ground for climate change*. Nairobi: Author.

Uzu, G., Schreck, E., Xiong, T., Macouin, M., Lévêque, T., Fayomi, B., et al. (2014). Urban market gardening in Africa: Foliar uptake of metal (loid) s and their bioaccessibility in vegetables; implications in terms of health risks. *Water, Air, & Soil Pollution, 225*, 1–13.

Van Veenhuizen, R. (2006). *Cities farming for the future: Urban agriculture for sustainable cities*. Leusden: RUAF Foundation, IDRC & IIPR.

Wang, J., Huang, B., Fu, D., & Atkinson, P. M. (2015). Spatiotemporal variation in surface urban heat island intensity and associated determinants across major Chinese cities. *Remote Sensing, 7*, 3670–3689.

Wang, J., Luo, Y., Teng, Y., Ma, W., Christie, P., & Li, Z. (2013). Soil contamination by phthalate esters in Chinese intensive vegetable production systems with different modes of use of plastic film. *Environmental Pollution, 180*, 265–273.

References

Waseem, A., Arshad, J., Iqbal, F., Sajjad, A., Mehmood, Z., & Murtaza, G. (2014). Pollution status of Pakistan: A retrospective review on heavy metal contamination of water, soil, and vegetables. *BioMed Research International, 2014*. https://doi.org/10.1155/2014/813206.

Wei, B., & Yang, L. (2010). A review of heavy metal contaminations in urban soils, urban road dusts and agricultural soils from China. *Microchemical Journal, 94*, 99–107.

WHO. (2014). *7 million premature deaths annually linked to air pollution*. Retrieved March 25, 2014, from http://www.who.int/mediacentre/news/releases/2014/air-pollution/en/

WHO. (2016). *Ambient (outdoor) air pollution in cities database 2014*. Retrieved August 09, 2016, from http://www.who.int/phe/health_topics/outdoorair/databases/cities/en/

WHO. (2017). *Household energy: Three billion left behind*. Retrieved November 16, 2017, from http://www.who.int/indoorair/publications/fflsection1.pdf

Xiao, R., Weng, Q., Ouyang, Z., Li, W., Schienke, E. W., & Zhang, Z. (2008). Land surface temperature variation and major factors in Beijing, China. *Photogrammetric Engineering & Remote Sensing, 74*, 451–461.

Yang, P., Mao, R., Shao, H., & Gao, Y. (2009). The spatial variability of heavy metal distribution in the suburban farmland of Taihang Piedmont Plain, China. *Comptes Rendus Biologies, 332*, 558–566.

Zhang, Y., Odeh, I. O. A., & Han, C. (2009). Bi-temporal characterization of land surface temperature in relation to impervious surface area, NDVI and NDBI, using a sub-pixel image analysis. *International Journal of Applied Earth Observation and Geoinformation, 11*, 256–264.

Zhong, S., Qian, Y., Zhao, C., Leung, R., & Yang, X. Q. (2015). A case study of urbanization impact on summer precipitation in the Greater Beijing Metropolitan Area: Urban heat island versus aerosol effects. *Journal of Geophysical Research: Atmospheres, 120*, 10903–10914.

Zhou, J., Hu, D., & Weng, Q. (2010). Analysis of radiation budget during the summer and winter in the metropolitan area of Beijing, China. *Journal of Applied Remote Sensing, 4*, 043513.

Chapter 7
The Three "E" Approach to Gender Mainstreaming in Climate Change: *Enumeration, Education, Empowerment*

Sustainable Development Goal 17: "Strengthen the means of implementation and revitalize the global partnership for sustainable development" (UN 2016a).

Introduction

The main objective of this book is to build a bridge between the vast body of scientific literature on climate change impacts and its disproportionate negative impacts on women and girls in the Global South. Through an in-depth review of existing literature and analysis of secondary data on gender indices and climate processes, an attempt has been made to highlight the importance of incorporating gender sensitive approaches to policy and adaptation strategies to climate change.

Through the previous chapters an effort has been made to reveal the unequal impacts of climate change on women and girls in the Global South. The general absence of gender-sensitive policies is partially due to the general lack of the collection of gender disaggregated data, and thus the absence of gender sensitive research. However, it is clear from the discussions in the previous chapters that gender equality is an integral element for an effective the long term adaptation and mitigation of the impacts of climate change. This is particularly critical in the Global South, which experience higher levels of gender inequalities combined with greater vulnerability to the impacts of climate change. Higher gender gaps in many of the countries in Global South, particularly Sub Saharan Africa and South Asia, place women and girls at a greater disadvantage vis a vis the impacts of climate change. In addition, due to lower levels of awareness among women and girls the overall wellbeing of the family is also compromised in many cases. Some of the main issues identified in the previous chapters are summarized below:

- There are substantial variations in the spatial patterns of gender gap and inequities in the Global South. The regions experiencing higher gender gaps are mostly concentrated in South Asia and Central Africa.

- The spatial concentrations of greater gender gaps overlap with the regions that are most vulnerable to climate change.
- Higher gender inequality in the Global South lead to women often having lesser access to resources leading to inadequate prior knowledge about an impending hazard due to reduced access to early warning messages and restrictions imposed by cultural and religious practices. The lack of empowerment and awareness restricts their opportunity to maintain a steady income from their main economic activities as a result of the adverse impacts of climate driven natural disasters.
- Additionally, societal norms and cultural practices result in limiting women's movements, which can be detrimental in the timely evacuation of women and girls when they are faced with a natural disaster. This can lead to not only physical harm, but also lingering emotional damage.
- As a result of the already occurring long term impacts of climate change, women and girls who are mostly responsible for collecting water and fuel wood in the rural areas are experiencing longer commute times. This makes them more vulnerable to harassments and exposes them to the harsher weather conditions such as high temperatures and water contamination.
- The role of climate change on human health is well documented. In this context women and girls are in general more vulnerable to the impacts of climate change, due to higher levels of malnutrition and exposure to unhealthy conditions such as non-hygienic living conditions and indoor air pollution from wood or coal burning stoves. Furthermore, in most of the traditional societies of the Global South, women and girls are responsible for taking care of those who are unwell, which exposes them to transmitted infections.
- In larger metropolitan areas of the Global South, due to growing rural urban migration that results in overcrowding in the cities and a clash of cultures, makes women and girls particularly vulnerable to harassment, as evidenced in the increasing instances of violence on women in major cities of the Global South.
- In addition, due to the overcrowding and declining yields in the rural and agricultural areas, the young men move to the big cities leaving behind the women to take care of the elderly and children under varying levels of food insecurity. The limited supply of essential food supplies lead to the widespread prevalence of malnutrition among women and girls, who often are the last to eat in the household.
- There is increasing evidence of the direct and indirect role of climate change on triggering conflicts in various regions of the Global South, such as Darfur conflict and the current Syrian conflict. As a result of these conflicts, vast number of populations are displaced from their homelands, and forced to move to new unfamiliar surroundings making them extremely vulnerable. Reports of violence on women and girls in the temporary settlements are very high, which are vastly under reported.
- Women have little or, in most cases, no participation in constructing formal social capital, including decision making processes, policies, and institutions, which deprive them when they require active support from social capital. In contrast, women have much larger influence in building informal social capital,

which immensely help the entire society, males included, towards addressing the aftermath of any climate-induced disaster.

Gender Mainstreaming

Therefore gender mainstreaming, which was established as a major global strategy for eliminating gender inequality in the fourth UN World Conference in Beijing in 1995, is essential for effectively addressing the gendered differences in the impacts of climate change in the Global South. Subsequently, the Economic and Social Council (ECOSOC) adopted a resolution on gender mainstreaming in 2001, to ensure that gender perspectives is taken into consideration in all of its work. The UN has also developed clear mandates for gender mainstreaming for all major areas of work of the UN. However, from the discussions in previous chapters, it is clear that we have a long way to go in terms of achieving gender equality specifically in the Global South. There is a critical need to address the role of social customs and polices at all levels to address gender equity issues in a more localized cultural context. From the discussion on some key issues in the previous chapters, it is clearly evident that women and girls are disadvantaged in terms of and against all socio-economic indicators. In addition they also constitute the majority of the poorest in the Global South, which coupled with to social customs makes it difficult for them to have equal access to resources or decision making units. There is also overwhelming evidence of women and girls working as unpaid caregivers, as well as being most vulnerable to gender based violence. Both of these aspects are essential to gender mainstreaming but cannot be addressed effectively due to inadequate data or information.

One of the most widely used definitions of gender mainstreaming is by the ECOSOC in 1997:

> Mainstreaming is the process of assessing the implications for women and men of any planned action, including legislation, policies or programs, in all areas and at all levels. It is a strategy for making women's as well as men's concerns and experiences an integral dimension of the design, implementation, monitoring and evaluation of policies and programs in all political, economic and societal spheres so that women and men benefit equally and inequality is not perpetuated. The ultimate goal is to achieve gender equality (ECOSOC 2016).

Gender mainstreaming constitutes a commitment to a comprehensive assessment of organizational structures, policies, and practices for gender bias.

In view of the prevalence of greater gender inequalities in the Global South, the differences in the impacts of climate change are also greater on women and girls. However, due to the lack of gender and age disaggregated data collection; there is very limited evidence to take effective actions.

Gender mainstreaming is increasingly being recognized as a major policy initiative by various governments and international organizations. However, the persisting patterns of gender inequalities in the Global South discussed in previous chapters highlight the need for greater initiatives toward gender mainstreaming. There needs to be a conscious effort at all levels to incorporate gender sensitive policies and more equitable involvement of women at all levels of decision making. In this regard, some of the specific measures that should be taken to promote gender mainstreaming resulting in greater equality for more effective adaptation to the impacts of climate change are outlined below briefly:

- The analysis of all potential and already occurring impacts of climate change and related policy formulations should take into consideration the gendered differences.
- Consideration of gender differences should become default for all policy formulations at all levels.
- Efforts should be made to narrow the differences in gender inequalities through greater opportunities for women and girls in education, empowerment, and overall wellbeing at all levels of decision making.
- Targeted interventions should be developed to create opportunities toward narrowing gender inequalities on a case by case basis.
- Gender mainstreaming should also be incorporated in analytical tasks which include the collection of gender and age disaggregated data. It is only with the collection of such data that we can accurately assess regional level gender gaps and address them.
- In view of the majority of the household and caregiving responsibilities with the women in the family, efforts should be made to somehow document their unpaid contributions in the society.

It is in this context the three "E" approach through which gender mainstreaming can be accomplished include Enumeration, Education, and Empowerment. Gender mainstreaming measures should start at a very young age and not just include girls but the entire community. The three sections below consist of an argument for the "E^3" approach for gender mainstreaming.

Enumeration

From the discussions in the previous chapters it is evident that the impacts of climate change are not gender neutral. However, one of the major obstacles in accurately assessing the impacts of climate change on women and girls is the absence of gender disaggregated data collection. As shown in the previous chapters, in the absence of such data it is often difficult to establish causality. Therefore, there is a vast gap in literature examining the gendered impacts of climate change on women and girls. Often the limited gender disaggregated data collected and available are not standardized or collected regularly across all regions. Thus it is difficult to do comparative analysis between various regions and gauge the change over time. As a result, there are very few scientific global scale studies on the gendered differences in the impacts of climate change on women and girls. However, the critical importance of these issues can be assessed from the numerous case studies at specific locations that have been conducted by various non-governmental and non-profit organizations, which include various agencies under the UN, World Resources Institute, and Global Gender and Climate Alliance. These case studies highlight the disparate impacts of climate change on women and girls, particularly in the Global South.

There are a wide variety of gender disaggregated data that are collected by various organizations on different parameters. However, most of the data collection is not systematic and widely enforced. As a result, it is often difficult to do a comparative and longitudinal analysis on gender gaps and inequalities. The estimated population of girls worldwide is 1.1 billion, which is almost 15% of the total world population (The Daily Star 2016). However, majority of the times their voices are not represented and they fall behind in terms of nutrition, health care, and education.

Therefore, not surprisingly in a recent meeting on International Day of the Girl on October 11, 2016 the UN focused on the theme *'Girls' Progress=Goals' Progress: A Global Data Movement' (UN* 2016b*)*. The meeting underscored the need for the collection of accurate, reliable, transparent, and comparable gender data to ensure progress for girls and leave no one behind for the 2030 agenda for sustainable development. The UN Secretary-General Ban Ki-Moon stressed "what cannot be measured cannot be managed. If we do not gather the data we need, we will never know if we are delivering on our promises." He further emphasized the importance of timely, high-quality data to identify areas which need attention. In addition the UN Women Executive Director, Phumzile Mlambo Ngcuka, also put emphasis on the importance of accurate, reliable, transparent and comparable gender data for ensuring the progress of girls. Thus, a new private public initiative was launched titled "Making Every Woman and Girl Count.' This program is jointly led by UN Women, the Government of Australia, the Bill and Melinda Gates Foundation, and Data2X (United Nations Foundation). It will address the need to increase the availability of accurate data on gender equality and women's rights to better informed policy formulation. This initiative will help generate gender sensitive data to ensure the results are used to shape policies and increase accountability. Currently less than 50 countries are able to provide gender and age disaggregated data, which makes it difficult to evaluate the impact of gender based violence, poverty, maternal deaths, and other critical issues (UN 2016b). Specifically, in 2012 approximately 17% of the 126 countries that responded to a survey by the UN Statistics Division had specific budget allocated toward gender statistics, 47% relied on an ad-hoc funds/project, and the remaining 39% had no funds at all (ECOSOC 2014). As a result, despite the awareness about issues that are holding back women and girls from a very young age, there is no extensive and systematic quantitative or qualitative data collection to conduct any evaluation. For instance, a global review of national gender statistics indicates the marginalization of the issue as a women's issue in various countries and thus not given much priority.

As a result of the absence of gender disaggregated data, there is also a general absence of publications focusing on this topic. It is also absent in publications of international agencies responsible for providing humanitarian aid in post disaster situations. The limited efforts in collecting such data are further complicated by the absence of no clear, widely-established standards for such data collection in the context of climate change.

Gender and age disaggregated quantitative data are helpful in generalizing information for a larger population about differences based on gender and age. Data collected over time can help in measuring progress made over time and the effectiveness of various policies implemented over time. Qualitative data on the other hand can provide individual or group level experiences of people based on their local culture, experiences, opinions, and attitudes. The methodology of qualitative data collection is mostly through unstructured interviews or focus groups. Often times, qualitative data can provide valuable information about personal experiences which may not be derived from mass scale quantitative information collected through structured surveys. In many cases, a structured survey may only capture the information from the male heads of the households but not the women and girls' perspectives. However, when girls and women are interviewed separately or in focus groups then they maybe more forthcoming and give a diverse perspective on the same situation. Efforts should be made to capture the views of both men and women in data collection. In addition to collecting gender and age disaggregated data, efforts should also be made to collect data on specific issues that affect either gender. Therefore a combination of both qualitative and quantitative data is important for addressing gender gaps and accurately assessing the impacts of climate change on women and girls at various spatial scales. Finally, the collection of age disaggregated data are also very important because often most data are aggregated on the basis of gender, which may not be able to address issues faced by adolescent girls versus women. Specifically, girls at a younger age have even less visibility and voice, while being the most vulnerable to the different impacts of climate change as highlighted in previous chapters.

Collection of big data should not only be limited to traditional methods and organizations, but also include innovative techniques such as social media and crowdsourcing. There is increasing awareness about the effective use of social media as a rich real time source of big data. Such data collection techniques through social media can remediate the biases and drawbacks of existing methods of data collection and prevent underreporting.

In addition to collecting data, there is also a need for timely analysis and presentation of the results for wider circulation and policy formulation. Effective visualization of the collected data (which includes mapping the gaps at various spatial scales and infographics) can be fast and effective for targeted initiatives by governments and nonprofit organizations at the local level. With reliable data and accurate analysis there can be targeted investment and efforts toward greater gender equality and awareness.

Education

One of the best investments for the overall progress of the society is in girls' education, health, and safety. Investing at an earlier age ensures that they grow up as more aware and empowered human beings, who can not only speak for themselves, but can also, care for their families. This ensures the overall long term wellbeing of the society. Free and equal access to education at all levels for girls of all economic and social status is essential for achieving gender equality, and therefore more informed adults in the future. It is estimated that an extra year in primary school can result in an increased adult wages of 10–20%, which increases to 15–25% as a result of an extra year in secondary school (UN 2016c). Furthermore, it is estimated that if all girls completed secondary education, then child mortality under five will be halved. With education comes awareness and the voice to speak up for themselves from a younger age, which can result in lesser number of child marriages and pregnancy at very young ages.

Furthermore, in the context of adapting to impacts of climate change, with better understanding of the basic science, they will grow up to be more informed citizens. Specifically, with projected increases in global population to 9.7 billion in 2050, most of this increase will take place in the Global South. This increase in population will be a mainly a result of a rapid decrease in mortality rates and a lower rate of decrease in fertility rates. Regionally, Africa's population is projected to increase by 1.2 billion people, concentrated mostly in countries such as Ethiopia, Nigeria, and the Democratic Republic of Congo (Kharas 2016). The existing infrastructure are already feeling the impacts of changing climate in the form of warmer temperatures, droughts, and extreme weather events resulting in large scale displacement of populations.

Moreover, these countries also rank very high in gender gap indices, showing high levels of gender inequalities for various indicators. Therefore the projected steep increase in population is unsustainable in the future. The impact of inverse correlation between education levels and fertility rate is widely validated. For instance, it is estimated that improved education in Africa will lead to 1.8 billion people less than the UN median variant suggested by 2100, as evidenced from the data analyzed from demographic and health surveys for several countries in Africa by the World Bank (Pradhan 2016). Lower population ultimately results in lowering carbon emissions.

In addition to malnutrition and malnourishment, food insecurity may also lead to lower levels of education among children. Based on the cultural norms in many of the countries in the Global South girls are forced to withdraw from primary and secondary schools to help with household chores such as taking care of younger siblings, collecting fuel-wood and water, and help in the agricultural sector. With the projected changes in climate processes, the amount of time the amount of time they spend doing these daily chores will only increase, and thus spend lesser time in school.

In this context, it is important to note that considerable progress has been made with regard to increase in enrolment of the girls in elementary and primary school with increased funding for education and favorable government policies. However, the dropout rates for girls in secondary schools are 467% compared to 35% for boys in Bangladesh, which is in the Global South (The Daily Star 2016).

The benefits of educating girls directly align with recent arguments about gender dividend, which is directly related to the more widely discussed demographic dividend. Gender dividend concentrated on increasing the volume of market (paid) work and the level of productivity of the female population. It argues for more productive and equitable economies by closing gender gaps in the labor market. Gender dividend can increase from lower fertility levels, because young educated women can focus more of their time in market paid work and raise their market productivity. Greater economic independence coupled with education can also lead to more investment in environmental friendly choices in the future.

Empowerment

It is also significant that most of the limited discourse on gender and climate change, as also highlighted in the previous chapters, is on the greater vulnerability of women and girls. While it is of utmost importance to highlight the greater vulnerability of women and girls in the Global South, which have very high prevalence of gender inequality and gaps. However, there is even less discussion about the role of women's empowerment in the adaptation and mitigation. The lack of power and limited access to financial resources puts women at a disadvantage to adapt to the impacts of climate change. Nevertheless, there is widespread evidence of the critical role of women in the conservation and reforestation efforts particularly in the rural areas of the Global South.

In many cultures some of the age old traditions with respect to lifestyle and traditions are passed through mothers to their daughters, which can be particularly useful for effective policies at the local level. This indigenous knowledge is extremely valuable for effective mitigation and adaptation to climate change impacts. The harnessing of indigenous knowledge can lead to more locally sustainable living conditions and utilizing traditions and customs which can help in mitigating some of the already occurring impacts of climate change.

In addition to enumeration and education, one of the logical results of the collection of gender disaggregated data, leading to better management and increased levels of education and awareness among women and girls, which will further lead to greater empowerment for women. Women's empowerment is defined as their ability to make strategic life choices where that ability was not present previously (Kabeer 1999). With the greater empowerment of women, there will be more increased representation of women in decision making roles. There will also be greater participation of women in policy development, which will result in more gender sensitive policies and awareness. For instance, without the participation of women in decision making processes, the decision to replace traditional crops may only serve the interests of men working in the fields and not address the problems faced by women. It is often the case that the existence of gender insensitive policies is due to the lack of awareness about gender issues. Thus, there will be better representation of gender issues at different levels. In addition, the participation of women in policies related to climate change mitigation can make them more sustainable and applicable at the local level while incorporating a gendered perspective.

Moreover, when women are involved as decision makers in the development of policies related to disasters and building resilience, then they are better able to adapt and manage the impacts of climate change. In this context, it is important to identify key pathways for women to gain empowerment, which can be bolstered by effective enumeration of gender gaps and greater education of girls from a young age.

Additionally, as described in previous chapters, local cultural norms can be restrictive for women and girls to protect themselves from climate hazards such as floods and hurricanes. Due to lower levels of training and awareness about various information and warning systems to protect themselves from such natural disasters, they are often not able to protect themselves or their dependents. By empowering them through awareness and greater decision making abilities, more lives can be saved in a timely and effective manner.

Conclusions

Thus in conclusion, the final recommendations for gender mainstreaming in the context of the projected and already occurring impacts of climate change are listed below:

- Data should be collected in a gender disaggregated form in order to adequately formulate gender sensitive adaptation and mitigation strategies. It will help us to identify the vulnerabilities and coping strategies for men and women separately.
- Ensuring the education of girls from a young age will lead to more well informed and aware adults, majority of whom will be in-charge of the future generation. Thus education of girls is a long term investment for future generations.
- There needs to be greater access to financial resources for women in the Global South, so that they can be more prepared for the impacts of climate change. It would also result in the reduction of losses and damages occurring as a result of climate change.
- More education and greater empowerment for women will help in lead to reducing their typically lower access to financial capital, which limit their ability to invest in measures that might have reduced their sensitivity and lack of preparedness to climate change impacts.
- All of this is possible once there are a concerted effort at all levels of decision making toward more gender sensitive policies. Gender equality need to be mainstreamed in all climate change related policies.

It is important to note that there is a vast body of literature examining the long term trends in climate change related processes. It is now important to link the results from the scientific literature to more gender sensitive impacts. In this regard there are increasing numbers of non-governmental organizations (NGOs) involved in the promotion of gender mainstreaming in various regions of the Global South. The results from these increasing numbers of initiatives on behalf of the various

NGOs supported by the results of scientific literature can help in creating awareness the gendered impacts of climate change, and strengthening the argument for more gender sensitive policies and data collection.

References

ECOSOC. (2014). *Using data to measure gender equality*. Retrieved November 4, 2014, from http://www.un.org/en/development/desa/news/gender/using-data-to-measure-gender-equality.html

ECOSOC. (2016). *Definition of gender mainstreaming*. Retrieved October 19, 2016, from http://www.ilo.org/public/english/bureau/gender/newsite2002/about/defin.htm

Kabeer, N. (1999). Resources, agency and achievements: Reflections on the measurement of women's empowerment. *Development and Change, 30*, 435–464.

Kharas, H. (2016). *Climate change, fertility and girl's education*. Retrieved February 16, 2016, from https://www.brookings.edu/blog/future-development/2016/02/16/climate-change-fertility-and-girls-education/

Pradhan, E. (2016). *Female education and childbearing: A closer look at the data*. Retrieved December 28, 2016, from http://blogs.worldbank.org/health/female-education-and-childbearing-closer-look-data

The Daily Star. (2016). *Girls' progress = Goals' progress: A global girl data movement*. Retrieved October 11, 2016, from http://www.thedailystar.net/supplements/unfpa-supplement/girls-progress-goals-progress-global-girl-data-movement-1297204

UN. (2016a). *SDGs: sustainable development knowledge platform. United Nations*. Retrieved December 29, 2016, from https://sustainabledevelopment.un.org/sdgs

UN. (2016b). *'All girls count,' says UN, calling for reliable data to uncover and tackle inequalities holding them back*. Retrieved October 11, 2016, from https://refugeesmigrants.un.org/%E2%80%98all-girls-count%E2%80%99-says-un-calling-reliable-data-uncover-and-tackle-inequalities-holding-them-back

UN. (2016c). *Inter-agency task force on rural women, fact sheet: Rural women and the millennium development goals and UNICEF (2011)*. Retrieved December 30, 2016, from http://www.unicef.org/media/media_58417.html

Index

A

Aerosols
anthropogenic sources, 11
global dimming and brightening, 11
mining and industrial activity, 11
natural sources, 11
substantial seasonal differences, spatial
patterns, 11
Age disaggregated data collection, 144
Air pollution
air quality levels, 126
in cities, 126
COPD, 129
DALY, 128, 129
deaths resulting from lower respiratory
infectious diseases, 127, 128
elevated levels, 125
greenhouse gas emissions, 125
health impacts, 126
on human health, 127
indoor, 128
inhaling smoke, 129
lower air quality, urban areas, 126
lower income populations, 125
outdoor, 128
particulate matter, 127
regional levels, 127
respiratory diseases, 126
spatial patterns of weighted population
exposure, 126, 127
UHI development, 125
urban levels, 125
urban sprawl and aerosol levels, 125
WHO, 125

Anthropogenic activities, 122
Anthropogenic processes, 126
Antiretroviral therapy, 47
Arid and semi-arid climates, 126

B

Beach erosion, 99
Beijing declaration in 1995, 27
Biodiversity hotspots, 133
Blood-stage malarial antigens, 62

C

Center for Disease Control (CDC), 62
Cholera, 87, 88
Chronic obstructive pulmonary disease
(COPD), 129
Civil conflicts, 109
Climate change and variability, 139
adverse social and economic impacts of, 2
anthropogenic activities, role of, 4
anthropogenic changes, 81
Atlantic Ocean and Indian Ocean, role in, 14
COP21 Conference in Paris, 1
desertification, 4
direct and immediate impacts, 5
earth's surface, 76
ENSO context, 14
fertility rate, 3–4
Fifth Assessment Report, IPCC, 2
freshwater resources, 78
gender mainstreaming (*see* Gender
mainstreaming)

© Springer International Publishing AG, part of Springer Nature 2018
S. Sen Roy, *Linking Gender to Climate Change Impacts in the Global South*,
Springer Climate, https://doi.org/10.1007/978-3-319-75777-3

149

150 Index

Climate change and variability (*cont.*)
　global average planetary temperatures, 1
　greenhouse gas emissions, 5
　heat waves, 2
　inter-annual and inter decadal scales, 5 (*see also* Long term trends)
　model predictions, 15–19
　monsoon rainfall in India, 14
　NCDC, 4
　policies development, 2
　precipitation, 81
　project models, 76
　regional scale long term trends, 4
　SAM, 14
　sea level rise, 4
　spatial scales, 2
　spatial variations, 2, 3
　SST, 14
　substantial decadal and inter-annual variability, 2
　vulnerable populations, 4
　water insecurity, 79
　water quality, 83
　water resources, 78
　water stress, 83
　weather events, 3, 12–13
Climate evacuee, 95
Climate-induced disaster, 141
Climate migrant, 95
Climate refugees
　classification, 96
　conflicts, 108–110
　cultural norms, 96
　definition, 93–95
　direct impacts, 93
　displacement, 93
　evacuee, 95
　extreme weather events (*see* Extreme weather events)
　food insecurity (*see* Food insecurity)
　forced displacement and migration of population, 96
　forced migrant, 95
　growing numbers and awareness, 95
　IDPs, 95
　indirect impacts, 93
　IPCC report, 93
　limited data, 94
　migrations, 95
　natural ecosystems, 93
　prevention and preparation, 95
　rising temperatures, 93
　sea level rise (*see* Sea level rise)

　statistics, 94, 95
　time scale, 94
　UNHCR, 94, 96
Climate variability, 67
Climate vulnerability index
　adaptive capacity, 48
　awareness, 50
　CGIAR, 50
　COP meetings, 50
　food security, 48
　GED, 50
　gendered differences, 51
　gendered implications, 48
　and gender inequalities/gaps, 48
　gender mainstreaming, 51
　gender sensitive climate policy, 50
　life-supporting, 48
　ND-GAIN, 48
　policy formulation, 51
　readiness index, 48–50
　spatial patterns, 49
　water availability, 48
　women's participation, 51
Cocoa production, 104
Communicable and non-communicable diseases, 53, 56
Conference of Parties (COP), 20, 50
Conflicts
　Christian Aid estimated, 108
　climate change to, 109
　Darfur, 109
　direct causality, 109
　ENSO, 109
　gender-based violence, 110
　global distribution, 108, 109
　indirect and direct impacts of climate change, 108
　and interpersonal violence, 110
　IPCC Report, 108
　mass exodus of population, 109
　propensity of, 109
　scarcity of resources, 109
　Syrian Crisis, 110–111
　and violence, 109
　violent conflicts, 108
　and warmer temperatures, 109
Consultative Group for International Agricultural Research (CGIAR), 50

D
Darfur conflict, 109, 140
Deforestation, Amazon rainforests, 9

Index   151

Democratic Republic of Congo (DRC), 87
Dengue
    CDC, 62
    climate conditions, 62, 64
    global teleconnections, 62
    pregnant women, 64
    transmission, 62
    virus, 64
Dengue fever, 62
Dengue virus, 64
Diarrheal diseases, 56, 87
Disability Adjusted Life Years (DALY), 128,
        129
Disaster-induced displacement, 101
Disasters, 101
Displacement, 93–96, 100–102, 108, 109
Diurnal temperature range (DTR), 6
Droughts
    climate change, 77
    evapotranspiration, 79
    floods, 81
    global south, 80
    length and intensity, 78
    livelihoods and draining, 76
    local subsistence crops, 76
    predictions, 85
    tropics and subtropics, 78
    women and girls, 79
Dumping grounds, 131

E
Economic and Social Council (ECOSOC), 141
Education, 39–42
    awareness and voice to speak up, 145
    benefits, 145
    and favorable government policies, 145
    food insecurity, 145
    free and equal access, 145
    gender dividend, 146
    GGGI
        bimodal distribution, 41
        economic and social equality, 39
        educational attainment ranks, 42
        equal access to educational
            opportunities, 39
        gender equality, 39
        indicators, 39, 40
        mean and expected years, 41
        schooling, 39, 41
        secondary education, 41, 42
        sub-indices, 39

        variables, 39
    global population, 145
    health and safety, 144
    high in gender gap, 145
    levels and fertility rate, 145
    malnourishment, 145
    malnutrition, 145
El Niño, 97
El Niño and Southern Oscillation (ENSO), 5,
        14, 16, 17, 65, 109
Empowerment, 146–147
    GGGI, 42–44
Enumeration
    collection of age disaggregated data, 144
    data collection, 144
    gender and age disaggregated data, 141, 142
    gender disaggregated data, 142, 143
    global review of national gender statistics,
        143
    International Day of the Girl on October
        11, 143
    Making Every Woman and Girl Count, 143
    non-governmental and non-profit
        organizations, 142–143
    organizations, 144
    qualitative data, 144
    quantitative data, 144
    social media and crowdsourcing, 144
    structured survey, 144
    timely analysis and presentation, 144
    UN Secretary-General Ban Ki-Moon, 143
    UN Women Executive Director, 143
    women and girls, 143
Extreme climate, 65
Extremes
    heavy precipitation events and hurricanes,
        4
    weather events, 12–13
Extreme weather events, 65
    additional costs due to flooding, 102
    complexity and frequency, 101
    disaster-induced displacement, 101
    disasters, 101
    displacements of population, 101
    floods and flashfloods, 101, 102

F
Female health indicators, 69–71
Female life expectancy, 45, 47
Food and Agricultural Organization (FAO),
        10, 29

152   Index

Food insecurity, 109
  agricultural activities, 130
  agricultural areas, 131
  agricultural sector, 103
  anthropogenic source of heavy metals, 130
  aquatic food supply, 103
  cash producing crops, 104
  climate change and variability, 103, 104
  collecting firewood, 105, 107
  contamination levels, 130
  crop yield modeling studies, 103
  dumping grounds, 131
  economies, 103
  female employment in agriculture sector,
     104, 105
  food shortage, 108
  food supply, 107
  Great Green Wall Initiative, 103, 104
  heavy metals, 129, 130
  high levels of malnutrition, 107
  higher food prices, 108
  labor force on rural farms, 131
  lesser-developed areas, 103
  levels of trace metal composition, 130
  limited access to credit, 106
  limited access to resources, 106
  livestock grazing, 103
  in lower latitudes, 104
  marine ecosystems, 104
  maternal health, 107
  model projections, 107
  prevalence of childhood stunting, 106, 107
  prevalence of female malnutrition, 106, 107
  quality and quantity, 103
  rainfall/heat stress, 103
  regional scale, 107
  resolve women's malnutrition, 108
  in rural areas, 104
  soil sample test, 130
  soil-crop system, 130
  urban agriculture, 131
  urban gardening, 130, 131
Floods
  in Bangladesh, 82
  climate change, 77
  drainage facilities, 83
  global exposure, 81
  and hurricanes, 87
  and inundation, 81
  natural disaster, 81, 87
  scarcity, 76
  South Asia, 75
  weather events, 78
Food shortage, 108
Forced migrant, 95

G
Gender and age disaggregated quantitative
    data, 144
Gender and development (GED), 50
Gender development index (GDI)
  command over economic resources, 33
  deviation of gender parity, 33
  education, 33
  health, 33
  rankings, 33, 34
Gender disaggregated data, 142, 143
Gender disparities, 45
Gender dividend, 146
Gendered differences, 31, 51
Gender equality, 36, 139
Gender gaps
  adverse impacts of climate change, 30
  Beijing declaration in 1995, 27
  CEDAW, 27
  changing climatic conditions, 30
  climate vulnerability index, 47–51
  decision-making processes, 29
  education, 39–42
  empowerment, 42–45
  GDI, 33
  gender equality and empower women, 28
  GGGI, 36–39
  GII, 34–36
  HDI, 31–32
  health and survival, 45–47
  indoor smoke, 30
  inequalities, 27–28
  mainstreaming methodology, 28
  Millennium Development Goals, 28
  negative impacts of climate change, 29
  rural areas women, 30
  sensitive approach, 29
  social and economic settings, 31
  UCLA's World Policy Analysis Center, 27
  UN agencies, 29
  unequal access to resources and decision-
    making processes, 30
  variability in climatic conditions, 29
  women's participation, 28
Gender inequalities, 31, 139, 141
Gender inequality index (GII)
  economic status, 35
  empowerment, 35
  and GDI rankings, 35
  human development costs, 34
  and lower HDI index, 34
  rankings across the Global South, 34, 35
  reproductive health, 35
Gender inequity, 31
Gender mainstreaming

Index

153

age disaggregated data, 142
  collection, 142
  definitions, 141
  direct and indirect role of climate change, 140
  ECOSOC, 141
  education, 144–146
  empowerment, 146–147
  enumeration, 142–144
  gender equality, 139
  gender inequality, 139
  gender-sensitive policies, 139
  metropolitan areas, 140
  organizational structures, policies and practices, 141
  policy formulations, 142
  policy initiative, 141
  role of climate change, 140
  societal norms and cultural practices, 140
  socio-economic indicators, 141
  spatial concentrations, 140
  targeted interventions, 142
  women and girls, 139, 141, 142
Gender sensitive climate policy, 50
Gender-sensitive policies, 139
General circulation models (GCMs), 15
Global gender gap index (GGGI)
  Asia Pacific region, 38
  average ranking, 38
  average temperatures, 37
  definition, 36
  economic development, 37
  economic participation and opportunity sub-index, 38
  education, 39–42
  empowerment, 42–45
  equality benchmark, 36
  extreme events, 37
  gender equality, 36
  Global Gender Gap Report, 36
  health and survival index, 38, 45–46
  higher rates, 38
  Latin America and Caribbean region, 38
  lowest ranks, 38
  male/female ratios, 36
  overall rank, 36, 37
  overall scores, 36, 37
Global Gender Gap Report, 36
Global South, 54–57
  changes, 77
  climate change, 82
  drinking water, 85

drought, 78
  droughts girls, 80
  flood risks, 83
  gender gaps, 77
  health of women and girls, 86
  natural disasters, 76
  populations, 77
  sanitation, 84, 85
  South and East Asia, 78
  tropical countries, 87
  water stressed areas, 79
  weather events, 75
  women and girls, 79
Global teleconnections, 68
Global warming, 93
Global warming hiatus, 7
Great Green Wall Initiative, 103, 104
Greenhouse gas emissions, 125
Greenness index, 124
Gross Domestic Product (GDP), 71
Ground water, 76
Gynecological problems, 69

H

Health
  climate variability, 67
  dengue, 62–65
  extreme climate and weather, 65
  GDP, 71
  heat waves, 66–67
  infectious diseases, 57–65
  malaria, 58–62
  out-of-pocket expenditures, 69
  WHO regions, 55
  women, 69
Health and survival, GGGI
  birth rate, 45
  causes of deaths, 47
  child mortality, 45
  female life expectancy, 47
  gender disparities, 45
  general agreement, 45
  global mortality rate, 47
  indicators, 45, 46
  male and female life expectancy (male-female), 46, 47
  MMR, 45, 47
  sex ratio, 45
  social norms, 45
Health care system, 77
Heat waves, 66–67

Index

Higher gender inequality, 140
Human Development Index (HDI), 32
  design, 31
  global scale, 31
  globalization, 31
  highlight people and capabilities, 31
  instability, 32
  lowest levels, 31
  natural processes, 32
  ranks, 32
  regional level changes, 31, 32
  secure and accessible, 32
  vulnerabilities, 32
Hurricanes, 54, 65
Hydrological cycle, 76, 78, 83

I
Incubation period, 57, 59
Infectious diseases
  classification, 57
  dengue fever, 62
  humid conditions, 58
  malaria, 58–62
  mosquito species, 58
  pathogens, 57
  temperatures, 58, 59
  WMO survey, 66
Inhaling smoke, 129
Insecticide-treated mosquito nets, 60
Inter-annual time scales, 67
Intergovernmental Panel on Climate Change
  (IPCC), 2
Internally Displaced Persons (IDPs), 95
International Day of the Girl on October 11,
  143

K
Katrina, 65

L
Land surface temperatures (LST), 123, 124
Land-use and land cover changes (LULCC), 7
  air quality and ambient temperatures, 125
  direct impacts, 125
  greenness index, 124
  by illegal squatter settlements, 123
  local level energy balance, 123
  LST, 123, 124
  NDVI, 124, 125

  surface temperatures, 124
  types, 123
  and UHI levels, 123
  unplanned urban sprawl, 123
  unplanned urbanization, 124
Long term trends
  average temperatures, 5
  Central America and Caribbean, 8
  central-southern Chile, 9
  climate forcings, 8
  data quality, 5
  DTR calculation, 6
  earth surface, 5
  evapotranspiration, 8
  FAO, 10
  gases and particulate matter, 10
  large metropolitan areas, 7
  long-term impacts, 8
  LULCC, 7
  NOAA, 6
  N_2O levels, 10
  pan evaporation, 9
  PDO, 5
  SACZ, 9
  surface air temperatures, 10
  surface-level specific humidity, 9
  temperatures, 8
  UHI effect, 7
  urbanization rate, 11
  WHO, urban population, 7

M
Making Every Woman and Girl Count, 143
Malaria, 58, 62
  age groups, 60
  cycle and temperature, 59
  deaths, 61
  insecticide-treated mosquito nets, 60
  spatial patterns, 58
  transmission, 59
Marine ecosystems, 104
Marshall islands, 18
Maternal mortality rates (MMR), 45–47
McKinsey Global Institute, 71
Melting ice caps, Mt. Kilimanjaro, 6
Menstrual hygiene management (MHM), 85
Middle East and North Africa (MENA), 78
Millennium Development Goals, 28
MODIS satellite imageries, 123, 124
Monsoon season, 75
Mosquito breeding, 62

Index 155

N

National Climate Data Center (NCDC), 4
National Oceanic and Atmospheric
    Administration (NOAA), 6
Natural ecosystems, 93
Non-communicable diseases, 69
Non-governmental organizations (NGOs), 147
Non-voluntary displacement of population, 95
Normalized Vegetation Index (NDVI), 124,
    125
North Atlantic Oscillation (NAO), 14
Notre Dame-Global Adaptation Index
    (ND-GAIN), 48

O

Out-of-pocket expenditures, 69

P

Pacific Access Category of migration, 100
Pacific Decadal Oscillation (PDO), 5
Pakistan floods, 94
Pan evaporation, 9
Personal security
    crime rates, 133
    large urban areas, 132
    metropolitan areas, 132
    sewage treatment, 132
    slums, 132, 133
    squatter settlements, 132
    transient population, 132
    urbanization, 133
    water mixes, 132
    women and girls, 132
Plasmodium falciparum, 58
Plasmodium vivax, 58
Policy formulation, 51
Policy making process, 51
Pollutants, continental transport, 12
Precipitation
    average annual temperature, 16
    ENSO, role of, 14
    extreme events, 12
    heavy events, 2
    and hurricanes, 4
    long-term impacts, 8
    mean sea level pressure difference, 14
    moisture convergence disruption, 8
    surface air temperatures, 8
    and wind speed, 17
Pre-monsoon heat wave, 66

Q

Qualitative data, 144
Quantitative data, 144

R

Readiness index, 49, 50
Resilient and sustainable cities
    air pollution, 125–129
    direct and indirect impacts, climate change,
        119
    employment and educational opportunities,
        118
    food insecurity, 129–131
    LULCC, 123–125
    personal security, 132–133
    resources and infrastructure development,
        119
    rural to urban migration, 118
    in slums, 119
    UHI, 120–123
    urban areas, 117
    urban population, 117, 118
Respiratory diseases, 126
Rising temperatures, 93
Rural to urban migration, 118

S

Sea level rise
    aggregate mining, 99
    anthropogenic activities, 99
    climate sensitive impacts, 100
    coastal flooding, 98
    costs of, 99
    densely-populated coastlines, 98
    direct effects, 99
    early adaptors, 99
    El Niño, 97
    extreme weather events, 100
    fluctuations in global climate, 97
    global mean sea levels, 98
    IPCC report, 98
    loading in river deltas, 97
    local economy, 98
    long-term sustained increases, 97
    loss of land to, 98, 99
    low-lying coastal communities, 98
    Pacific Access Category of migration, 100
    rate of urbanization and industrialization,
        97
    regional level, 100

156   Index

Sea level rise (*cont.*)
  salinity levels, 98
  short-term increases, 97
  SIDS, 98, 99
  subsidence due to sediment compaction, 97
  subsiding coastal megacities, 97, 98
Seasonal climate variability, 68
Seasonal rainfall, 75
Sea-surface temperatures (SST), 14
Sewage treatment, 132
Slums
  growths, 119
  low lying areas of cities, 132
  people living in, 119
  population, 132
  and squatters, 130–132
  urban population living in, 132, 133
  weak materials, 132
Small island developing states (SIDS), 98, 99
Socio-economic development, 57
Socio-economic indicators, 141
Soil-crop system, 130
Soil moisture, 76
South Atlantic Convergence Zone (SACZ), 9
South Kivu, 87
South Pacific island of Vanuatu, 76
Southern Annular Mode (SAM), 14
Stern Review Report, 94
Subsiding coastal megacities, 97, 98
Sustainable Development Goals (SDGs), 53
Syrian conflict, 140
Syrian crisis, 110

T
Teleconnection pattern calculation, 13
Temperatures
  combined land and ocean surface, 2
  GHGs emissions, 10
  global average planetary, 1
  long term trends, 5
  NCDC, 4
  and precipitation, 8
Torrential rains, 75
Transient population, 132

U
UCLA's World Policy Analysis Center, 27
UN Convention on the Elimination of All
      Forms of Discrimination against
      Women (CEDAW), 27
United Nations Development Program
      (UNDP), 29

United Nations Educational, Scientific and
      Cultural Organization (UNESCO),
      29
UN Secretary-General Ban Ki-Moon, 143
UN Women Executive Director, 143
Urban areas, 117
Urban expansion, 117
Urban gardening, 130, 131
Urban heat island (UHI)
  administrative/financial capitals, 122
  analysis, 122
  anthropogenic activities, 122
  characterization, 120, 122
  definition, 120
  densification of urban areas, 120
  development, 120, 122
  IPCC report, 120
  and LULCC, 121
  population in urban agglomerations, 120,
      121
  rate of urbanization, 121
  role of urbanization, 123
  substantial spatial and temporal variations,
      122
  in summer season, 122
  surface level, 122
  in temperate mid-latitude cities, 120
  urban agglomerations, 122
  urban expansion, 121
Urban population, 117, 118

V
Vector borne diseases, 57
Vector-borne infective agent, 57
Violent conflicts, 108

W
Warmest period, 65
Water-borne diseases
  cholera epidemics, 87
  contamination, 86
  drinking water, 86
  empowerment, 86
  female death rates, 87
  flooding, 87
  hospitalization rates, 87
  infectious disease, 86
  natural disasters, 86, 87
  poor sanitation, 86
  tropical countries, 87
  washing and cleaning, 87
  women's health, 87

Index

157

Water quality
  availability and distribution, 76
  climate change, 83
  drinking water, 84
  gender inequality indices, 84
  global rural population, 85
  global water demand, 83
  grassroots level, 86
  household environment, 85
  hydrological cycle, 83
  hygiene, 85
  measurement, 86
  MHM, 85
  sanitation, 84, 85
  scarcity, 85
  shared sanitation, 85
  in Sub Saharan Africa, 85
  tanker-trucks/carts, 84
  temperature, 83
Water supply, 78, 83, 86
Weather-related disasters, 101
WHO/UNICEF Joint Monitoring Programmed
      for Water Supply and Sanitation
      2015, 83
Women and girls, 94–96, 101, 102, 107, 108,
      110, 119, 128, 132, 133, 139–144,
      146, 147
World Health Organization (WHO), 29, 54,
      125
World Map, 55
World Meteorological Organization (WMO),
      66